McGraw-Hill
Mathematics

Power ⚡ Facts!
Workbook

5

McGraw-Hill
School Division

New York   Farmington

McGraw-Hill School Division ☄

A Division of The **McGraw·Hill** *Companies*

Copyright © McGraw-Hill School Division,
a Division of the Educational and Professional Publishing Group of The McGraw-Hill Companies, Inc.
All rights reserved.

McGraw-Hill School Division
Two Penn Plaza
New York, New York 10121-2298

Printed in the United States of America

ISBN 0-02-100505-2

1 2 3 4 5 6 7 8 9 024 05 04 03 02 01 00

GRADE 5
Contents

Name_____

2 + 3 = 5

$$\begin{array}{r} 4 \\ +2 \\ \hline 6 \end{array}$$

Add. Write the sum.

A 2 + 1 = ___   3 + 2 = ___   1 + 3 = ___

B 1 + 4 = ___   3 + 0 = ___   4 + 1 = ___

C 2 + 4 = ___   0 + 4 = ___   5 + 1 = ___

D 0 + 1 = ___   3 + 3 = ___   1 + 2 = ___

E
$$\begin{array}{r} 2 \\ +3 \\ \hline \end{array} \quad \begin{array}{r} 1 \\ +2 \\ \hline \end{array} \quad \begin{array}{r} 0 \\ +2 \\ \hline \end{array} \quad \begin{array}{r} 3 \\ +1 \\ \hline \end{array} \quad \begin{array}{r} 4 \\ +0 \\ \hline \end{array} \quad \begin{array}{r} 1 \\ +1 \\ \hline \end{array}$$

F
$$\begin{array}{r} 0 \\ +0 \\ \hline \end{array} \quad \begin{array}{r} 5 \\ +0 \\ \hline \end{array} \quad \begin{array}{r} 4 \\ +2 \\ \hline \end{array} \quad \begin{array}{r} 1 \\ +5 \\ \hline \end{array} \quad \begin{array}{r} 0 \\ +6 \\ \hline \end{array} \quad \begin{array}{r} 2 \\ +2 \\ \hline \end{array}$$

G
$$\begin{array}{r} 1 \\ +0 \\ \hline \end{array} \quad \begin{array}{r} 0 \\ +5 \\ \hline \end{array} \quad \begin{array}{r} 3 \\ +3 \\ \hline \end{array} \quad \begin{array}{r} 6 \\ +0 \\ \hline \end{array} \quad \begin{array}{r} 0 \\ +3 \\ \hline \end{array} \quad \begin{array}{r} 2 \\ +0 \\ \hline \end{array}$$

Name_____

5 + 1 = 6

$$\begin{array}{r} 6 \\ +2 \\ \hline 8 \end{array}$$

Add. Write the sum.

A   7 + 1 = _8_   3 + 5 = _8_   7 + 1 = _8_   5 + 3 = _8_

B   2 + 6 = _8_   1 + 6 = _1_   5 + 2 = _6_   ⟨6 + 2 = _7_⟩

C   4 + 3 = _1_   ⟨0 + 7 = _0_⟩   3 + 4 = _1_   4 + 4 = _8_

D   8 + 0 = _8_   2 + 5 = _1_   ⟨0 + 8 = _0_⟩   1 + 7 = _8_

E
$$\begin{array}{r} 4 \\ +3 \\ \hline 7 \end{array} \qquad \begin{array}{r} 0 \\ +7 \\ \hline 0 \end{array} \qquad \begin{array}{r} 1 \\ +6 \\ \hline 7 \end{array} \qquad \begin{array}{r} 3 \\ +5 \\ \hline 8 \end{array} \qquad \begin{array}{r} 6 \\ +2 \\ \hline 8 \end{array} \qquad \begin{array}{r} 8 \\ +0 \\ \hline 8 \end{array}$$

F
$$\begin{array}{r} 4 \\ +4 \\ \hline 8 \end{array} \qquad \begin{array}{r} 2 \\ +5 \\ \hline 6 \end{array} \qquad \begin{array}{r} 0 \\ +8 \\ \hline 0 \end{array} \qquad \begin{array}{r} 7 \\ +1 \\ \hline 8 \end{array} \qquad \begin{array}{r} 3 \\ +4 \\ \hline 7 \end{array} \qquad \begin{array}{r} 5 \\ +3 \\ \hline 8 \end{array}$$

G
$$\begin{array}{r} 7 \\ +0 \\ \hline 7 \end{array} \qquad \begin{array}{r} 1 \\ +7 \\ \hline 8 \end{array} \qquad \begin{array}{r} 5 \\ +2 \\ \hline 7 \end{array} \qquad \begin{array}{r} 3 \\ +4 \\ \hline 7 \end{array} \qquad \begin{array}{r} 6 \\ +1 \\ \hline 7 \end{array} \qquad \begin{array}{r} 2 \\ +6 \\ \hline 8 \end{array}$$

Name_____

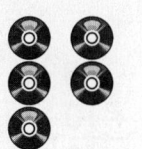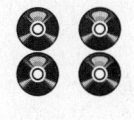

5 + 4 = 9

$$\begin{array}{r} 7 \\ +5 \\ \hline 12 \end{array}$$

Add. Write the sum.

A 3 + 7 = 10   4 + 6 = 10   5 + 7 = 12   9 + 3 = 12

B 6 + 5 = 11   8 + 2 = 10   7 + 3 = 10   9 + 1 = 10

C 1 + 8 = 9   4 + 7 = 11   6 + 3 = 9   2 + 9 = 11

D 6 + 4 = 10   3 + 8 = 11   8 + 4 = 12   5 + 6 = 11

E

$$\begin{array}{r} 7 \\ +4 \\ \hline 11 \end{array} \qquad \begin{array}{r} 3 \\ +6 \\ \hline 9 \end{array} \qquad \begin{array}{r} 5 \\ +5 \\ \hline 10 \end{array} \qquad \begin{array}{r} 10 \\ +2 \\ \hline 12 \end{array} \qquad \begin{array}{r} 7 \\ +5 \\ \hline 12 \end{array} \qquad \begin{array}{r} 5 \\ +4 \\ \hline 9 \end{array}$$

F

$$\begin{array}{r} 4 \\ +5 \\ \hline 9 \end{array} \qquad \begin{array}{r} 1 \\ +9 \\ \hline 10 \end{array} \qquad \begin{array}{r} 8 \\ +3 \\ \hline 11 \end{array} \qquad \begin{array}{r} 1 \\ +10 \\ \hline 11 \end{array} \qquad \begin{array}{r} 8 \\ +1 \\ \hline 9 \end{array} \qquad \begin{array}{r} 9 \\ +0 \\ \hline 9 \end{array}$$

G

$$\begin{array}{r} 2 \\ +7 \\ \hline 9 \end{array} \qquad \begin{array}{r} 0 \\ +9 \\ \hline 9 \end{array} \qquad \begin{array}{r} 10 \\ +1 \\ \hline 11 \end{array} \qquad \begin{array}{r} 2 \\ +8 \\ \hline 10 \end{array} \qquad \begin{array}{r} 4 \\ +8 \\ \hline 12 \end{array} \qquad \begin{array}{r} 0 \\ +10 \\ \hline 10 \end{array}$$

H

$$\begin{array}{r} 7 \\ +2 \\ \hline 9 \end{array} \qquad \begin{array}{r} 10 \\ +0 \\ \hline 10 \end{array} \qquad \begin{array}{r} 3 \\ +9 \\ \hline 12 \end{array} \qquad \begin{array}{r} 2 \\ +10 \\ \hline 12 \end{array} \qquad \begin{array}{r} 9 \\ +2 \\ \hline 11 \end{array} \qquad \begin{array}{r} 6 \\ +6 \\ \hline 12 \end{array}$$

Name_____

7 + 9 = 16

$$\begin{array}{r} 8 \\ +10 \\ \hline 18 \end{array}$$

Add. Write the sum.

A 9 + 4 = 13   8 + 9 = 17   9 + 8 = 17   9 + 10 = 19

B 8 + 6 = 14   9 + 7 = 16   6 + 8 = 14   5 + 10 = 15

C 9 + 9 = 18   5 + 8 = 13   4 + 9 = 13   10 + 8 = 18

D
$$\begin{array}{r} 6 \\ +7 \\ \hline 13 \end{array} \quad \begin{array}{r} 10 \\ +6 \\ \hline 16 \end{array} \quad \begin{array}{r} 9 \\ +6 \\ \hline 15 \end{array} \quad \begin{array}{r} 6 \\ +10 \\ \hline 16 \end{array} \quad \begin{array}{r} 8 \\ +8 \\ \hline 16 \end{array} \quad \begin{array}{r} 3 \\ +10 \\ \hline 13 \end{array}$$

E
$$\begin{array}{r} 9 \\ +5 \\ \hline 14 \end{array} \quad \begin{array}{r} 10 \\ +5 \\ \hline 15 \end{array} \quad \begin{array}{r} 8 \\ +10 \\ \hline 18 \end{array} \quad \begin{array}{r} 7 \\ +9 \\ \hline 16 \end{array} \quad \begin{array}{r} 4 \\ +10 \\ \hline 14 \end{array} \quad \begin{array}{r} 6 \\ +9 \\ \hline 15 \end{array}$$

F
$$\begin{array}{r} 8 \\ +5 \\ \hline 13 \end{array} \quad \begin{array}{r} 7 \\ +10 \\ \hline 17 \end{array} \quad \begin{array}{r} 8 \\ +7 \\ \hline 15 \end{array} \quad \begin{array}{r} 7 \\ +7 \\ \hline 14 \end{array} \quad \begin{array}{r} 10 \\ +4 \\ \hline 14 \end{array} \quad \begin{array}{r} 10 \\ +10 \\ \hline 20 \end{array}$$

G
$$\begin{array}{r} 5 \\ +9 \\ \hline 14 \end{array} \quad \begin{array}{r} 7 \\ +6 \\ \hline 13 \end{array} \quad \begin{array}{r} 10 \\ +9 \\ \hline 19 \end{array} \quad \begin{array}{r} 7 \\ +8 \\ \hline 15 \end{array} \quad \begin{array}{r} 10 \\ +3 \\ \hline 13 \end{array} \quad \begin{array}{r} 10 \\ +7 \\ \hline 17 \end{array}$$

The sum of 0 and a number is the number.

$$2 + 0 = 2$$
$$0 + 2 = 2$$

$$\begin{array}{r} 6 \\ +\,0 \\ \hline 6 \end{array}$$

$$\begin{array}{r} 0 \\ +\,6 \\ \hline 6 \end{array}$$

Add. Write the sum.

A $8 + 0 = \underline{8}$ $6 + 1 = \underline{7}$ $0 + 2 = \underline{2}$

B $4 + 0 = \underline{4}$ $0 + 3 = \underline{3}$ $3 + 1 = \underline{4}$

C $0 + 6 = \underline{6}$ $4 + 4 = \underline{8}$ $9 + 0 = \underline{9}$

D $1 + 1 = \underline{2}$ $2 + 0 = \underline{2}$ $0 + 5 = \underline{5}$

E
$$\begin{array}{r} 0 \\ +\,4 \\ \hline 4 \end{array} \qquad \begin{array}{r} 3 \\ +\,0 \\ \hline 3 \end{array} \qquad \begin{array}{r} 0 \\ +\,8 \\ \hline 8 \end{array} \qquad \begin{array}{r} 5 \\ +\,5 \\ \hline 10 \end{array} \qquad \begin{array}{r} 0 \\ +\,9 \\ \hline 0 \end{array}$$

F
$$\begin{array}{r} 0 \\ +\,3 \\ \hline 3 \end{array} \qquad \begin{array}{r} 1 \\ +\,8 \\ \hline 9 \end{array} \qquad \begin{array}{r} 5 \\ +\,0 \\ \hline 5 \end{array} \qquad \begin{array}{r} 0 \\ +\,6 \\ \hline 6 \end{array} \qquad \begin{array}{r} 0 \\ +\,7 \\ \hline 7 \end{array}$$

G
$$\begin{array}{r} 4 \\ +\,0 \\ \hline 4 \end{array} \qquad \begin{array}{r} 2 \\ +\,1 \\ \hline 3 \end{array} \qquad \begin{array}{r} 0 \\ +\,0 \\ \hline 0 \end{array} \qquad \begin{array}{r} 1 \\ +\,0 \\ \hline 1 \end{array} \qquad \begin{array}{r} 7 \\ +\,1 \\ \hline 8 \end{array}$$

Name_____

You can add a greater number to a lesser number.

4 + 5 = 9

$$\begin{array}{r} 6 \\ + 8 \\ \hline 14 \end{array}$$

Add. Write the sum.

A 4 + 9 = ___   3 + 8 = ___   3 + 5 = ___   1 + 10 = ___

B 2 + 5 = ___   4 + 8 = ___   6 + 9 = ___   8 + 10 = ___

C 6 + 7 = ___   1 + 3 = ___   7 + 8 = ___   3 + 10 = ___

D
$$\begin{array}{r} 1 \\ + 2 \\ \hline \end{array} \qquad \begin{array}{r} 4 \\ + 5 \\ \hline \end{array} \qquad \begin{array}{r} 2 \\ + 6 \\ \hline \end{array} \qquad \begin{array}{r} 10 \\ + 8 \\ \hline \end{array} \qquad \begin{array}{r} 2 \\ + 9 \\ \hline \end{array} \qquad \begin{array}{r} 6 \\ + 10 \\ \hline \end{array}$$

E
$$\begin{array}{r} 5 \\ + 6 \\ \hline \end{array} \qquad \begin{array}{r} 7 \\ + 10 \\ \hline \end{array} \qquad \begin{array}{r} 2 \\ + 4 \\ \hline \end{array} \qquad \begin{array}{r} 5 \\ + 7 \\ \hline \end{array} \qquad \begin{array}{r} 7 \\ + 9 \\ \hline \end{array} \qquad \begin{array}{r} 9 \\ + 10 \\ \hline \end{array}$$

F
$$\begin{array}{r} 1 \\ + 4 \\ \hline \end{array} \qquad \begin{array}{r} 3 \\ + 6 \\ \hline \end{array} \qquad \begin{array}{r} 5 \\ + 9 \\ \hline \end{array} \qquad \begin{array}{r} 3 \\ + 5 \\ \hline \end{array} \qquad \begin{array}{r} 2 \\ + 10 \\ \hline \end{array} \qquad \begin{array}{r} 4 \\ + 7 \\ \hline \end{array}$$

G
$$\begin{array}{r} 3 \\ + 7 \\ \hline \end{array} \qquad \begin{array}{r} 5 \\ + 10 \\ \hline \end{array} \qquad \begin{array}{r} 2 \\ + 8 \\ \hline \end{array} \qquad \begin{array}{r} 3 \\ + 4 \\ \hline \end{array} \qquad \begin{array}{r} 1 \\ + 6 \\ \hline \end{array} \qquad \begin{array}{r} 4 \\ + 10 \\ \hline \end{array}$$

Name_____

You can add a lesser number to a greater number.

6 + 2 = 8

$\begin{array}{r} 9 \\ +5 \\ \hline 14 \end{array}$

Add. Write the sum.

A   9 + 1 = ___   6 + 5 = ___   10 + 2 = ___   4 + 3 = ___

B   7 + 2 = ___   2 + 1 = ___   9 + 4 = ___   6 + 4 = ___

C   8 + 3 = ___   7 + 5 = ___   10 + 9 = ___   9 + 8 = ___

D   5 + 3 = ___   9 + 4 = ___   7 + 1 = ___   8 + 5 = ___

E
$\begin{array}{r} 10 \\ +8 \\ \hline \end{array}$
$\begin{array}{r} 9 \\ +5 \\ \hline \end{array}$
$\begin{array}{r} 6 \\ +5 \\ \hline \end{array}$
$\begin{array}{r} 8 \\ +6 \\ \hline \end{array}$
$\begin{array}{r} 6 \\ +1 \\ \hline \end{array}$
$\begin{array}{r} 10 \\ +6 \\ \hline \end{array}$

F
$\begin{array}{r} 8 \\ +5 \\ \hline \end{array}$
$\begin{array}{r} 10 \\ +5 \\ \hline \end{array}$
$\begin{array}{r} 9 \\ +7 \\ \hline \end{array}$
$\begin{array}{r} 8 \\ +1 \\ \hline \end{array}$
$\begin{array}{r} 9 \\ +6 \\ \hline \end{array}$
$\begin{array}{r} 8 \\ +7 \\ \hline \end{array}$

G
$\begin{array}{r} 5 \\ +4 \\ \hline \end{array}$
$\begin{array}{r} 7 \\ +6 \\ \hline \end{array}$
$\begin{array}{r} 8 \\ +4 \\ \hline \end{array}$
$\begin{array}{r} 3 \\ +2 \\ \hline \end{array}$
$\begin{array}{r} 5 \\ +1 \\ \hline \end{array}$
$\begin{array}{r} 9 \\ +3 \\ \hline \end{array}$

Name_____

Count on to add. Use a number line.

Say 8. Then count 9, 10, 11.

$$\begin{array}{r} 8 \\ +3 \\ \hline 12 \end{array}$$

8 + 3 = 11

0 1 2 3 4 5 6 7 8 9 10 11 12 13 14 15 16 17 18 19 20

Write the sum.

A 3 + 2 = ___   6 + 4 = ___   8 + 3 = ___   5 + 2 = ___

B 6 + 3 = ___   9 + 2 = ___   7 + 3 = ___   10 + 4 = ___

C 9 + 4 = ___   10 + 2 = ___   5 + 3 = ___   9 + 3 = ___

D
$$\begin{array}{r} 7 \\ +1 \\ \hline \end{array}$$
$$\begin{array}{r} 5 \\ +4 \\ \hline \end{array}$$
$$\begin{array}{r} 10 \\ +3 \\ \hline \end{array}$$
$$\begin{array}{r} 6 \\ +2 \\ \hline \end{array}$$
$$\begin{array}{r} 3 \\ +4 \\ \hline \end{array}$$

E
$$\begin{array}{r} 10 \\ +1 \\ \hline \end{array}$$
$$\begin{array}{r} 3 \\ +3 \\ \hline \end{array}$$
$$\begin{array}{r} 7 \\ +4 \\ \hline \end{array}$$
$$\begin{array}{r} 4 \\ +2 \\ \hline \end{array}$$
$$\begin{array}{r} 6 \\ +1 \\ \hline \end{array}$$

F
$$\begin{array}{r} 5 \\ +3 \\ \hline \end{array}$$
$$\begin{array}{r} 10 \\ +2 \\ \hline \end{array}$$
$$\begin{array}{r} 9 \\ +1 \\ \hline \end{array}$$
$$\begin{array}{r} 2 \\ +4 \\ \hline \end{array}$$
$$\begin{array}{r} 3 \\ +2 \\ \hline \end{array}$$

G
$$\begin{array}{r} 10 \\ +4 \\ \hline \end{array}$$
$$\begin{array}{r} 8 \\ +4 \\ \hline \end{array}$$
$$\begin{array}{r} 6 \\ +3 \\ \hline \end{array}$$
$$\begin{array}{r} 9 \\ +2 \\ \hline \end{array}$$
$$\begin{array}{r} 4 \\ +4 \\ \hline \end{array}$$

Use the sum of a double to add a double plus one.

$$\begin{array}{r} 4 \\ +4 \\ \hline 8 \end{array} \qquad \begin{array}{r} 4 \\ +5 \\ \hline 9 \end{array} \qquad \begin{array}{r} 8 \\ +8 \\ \hline 16 \end{array} \qquad \begin{array}{r} 8 \\ +9 \\ \hline 17 \end{array}$$

Add. Write the sum.

A $3 + 3 = \underline{6}$ $3 + 4 = \underline{7}$ **B** $5 + 5 = \underline{10}$ $5 + 6 = \underline{11}$

C $0 + 0 = \underline{0}$ $0 + 1 = \underline{1}$ **D** $7 + 7 = \underline{14}$ $7 + 8 = \underline{15}$

E $2 + 2 = \underline{4}$ $2 + 3 = \underline{5}$ **F** $6 + 6 = \underline{12}$ $6 + 7 = \underline{13}$

G $8 + 8 = \underline{16}$ $8 + 9 = \underline{17}$ **H** $3 + 3 = \underline{6}$ $3 + 4 = \underline{7}$

I $\begin{array}{r} 2 \\ +2 \\ \hline 4 \end{array}$ $\begin{array}{r} 2 \\ +3 \\ \hline 5 \end{array}$ **J** $\begin{array}{r} 6 \\ +6 \\ \hline 12 \end{array}$ $\begin{array}{r} 6 \\ +7 \\ \hline 13 \end{array}$ **K** $\begin{array}{r} 4 \\ +4 \\ \hline 8 \end{array}$ $\begin{array}{r} 4 \\ +5 \\ \hline 9 \end{array}$

L $\begin{array}{r} 1 \\ +1 \\ \hline 2 \end{array}$ $\begin{array}{r} 1 \\ +2 \\ \hline 3 \end{array}$ **M** $\begin{array}{r} 8 \\ +8 \\ \hline 16 \end{array}$ $\begin{array}{r} 8 \\ +9 \\ \hline 17 \end{array}$ **N** $\begin{array}{r} 5 \\ +5 \\ \hline 10 \end{array}$ $\begin{array}{r} 5 \\ +6 \\ \hline 11 \end{array}$

O $\begin{array}{r} 7 \\ +7 \\ \hline 14 \end{array}$ $\begin{array}{r} 7 \\ +8 \\ \hline 15 \end{array}$ **P** $\begin{array}{r} 3 \\ +3 \\ \hline 6 \end{array}$ $\begin{array}{r} 3 \\ +4 \\ \hline 7 \end{array}$ **Q** $\begin{array}{r} 0 \\ +0 \\ \hline 0 \end{array}$ $\begin{array}{r} 0 \\ +1 \\ \hline 1 \end{array}$

Make a ten. Then add on the rest of the second number.

$$\begin{array}{r} 7 \\ +6 \\ \hline 13 \end{array}$$

$7 + 6 = 13$

Say $7 + 3 = 10$ and $6 - 3 = 3$
$7 + 6$ equals $10 + 3$.
$10 + 3 = 13$, so $7 + 6 = 13$.

Make a ten. Then write the sum.

A $8 + 3 =$ ___ $4 + 8 =$ ___ $7 + 6 =$ ___ $9 + 5 =$ ___

B $8 + 5 =$ ___ $5 + 6 =$ ___ $5 + 9 =$ ___ $3 + 8 =$ ___

C $8 + 7 =$ ___ $8 + 8 =$ ___ $7 + 7 =$ ___ $7 + 4 =$ ___

D
$$\begin{array}{r} 7 \\ +7 \\ \hline \end{array} \qquad \begin{array}{r} 6 \\ +5 \\ \hline \end{array} \qquad \begin{array}{r} 8 \\ +7 \\ \hline \end{array} \qquad \begin{array}{r} 6 \\ +9 \\ \hline \end{array} \qquad \begin{array}{r} 6 \\ +7 \\ \hline \end{array}$$

E
$$\begin{array}{r} 3 \\ +9 \\ \hline \end{array} \qquad \begin{array}{r} 9 \\ +9 \\ \hline \end{array} \qquad \begin{array}{r} 6 \\ +8 \\ \hline \end{array} \qquad \begin{array}{r} 5 \\ +7 \\ \hline \end{array} \qquad \begin{array}{r} 9 \\ +4 \\ \hline \end{array}$$

F
$$\begin{array}{r} 8 \\ +5 \\ \hline \end{array} \qquad \begin{array}{r} 8 \\ +8 \\ \hline \end{array} \qquad \begin{array}{r} 5 \\ +9 \\ \hline \end{array} \qquad \begin{array}{r} 9 \\ +2 \\ \hline \end{array} \qquad \begin{array}{r} 6 \\ +6 \\ \hline \end{array}$$

G
$$\begin{array}{r} 8 \\ +4 \\ \hline \end{array} \qquad \begin{array}{r} 6 \\ +3 \\ \hline \end{array} \qquad \begin{array}{r} 8 \\ +3 \\ \hline \end{array} \qquad \begin{array}{r} 9 \\ +6 \\ \hline \end{array} \qquad \begin{array}{r} 3 \\ +8 \\ \hline \end{array}$$

An addition fact can be turned around. The sum is the same.

$$\begin{array}{r} 5 \\ +3 \\ \hline 8 \end{array} \qquad \begin{array}{r} 3 \\ +5 \\ \hline 8 \end{array}$$

$$5 + 7 = 12$$
$$7 + 5 = 12$$

Add. Write the sum.

A $1 + 4 =$ ___ $4 + 1 =$ ___   **B** $6 + 2 =$ ___ $2 + 6 =$ ___

C $5 + 2 =$ ___ $2 + 5 =$ ___   **D** $7 + 1 =$ ___ $1 + 7 =$ ___

E $3 + 6 =$ ___ $6 + 3 =$ ___   **F** $8 + 2 =$ ___ $2 + 8 =$ ___

G $9 + 4 =$ ___ $4 + 9 =$ ___   **H** $7 + 5 =$ ___ $5 + 7 =$ ___

I $\begin{array}{r} 0 \\ +9 \\ \hline \end{array}$ $\begin{array}{r} 9 \\ +0 \\ \hline \end{array}$ **J** $\begin{array}{r} 7 \\ +4 \\ \hline \end{array}$ $\begin{array}{r} 4 \\ +7 \\ \hline \end{array}$ **K** $\begin{array}{r} 5 \\ +3 \\ \hline \end{array}$ $\begin{array}{r} 3 \\ +5 \\ \hline \end{array}$

L $\begin{array}{r} 8 \\ +4 \\ \hline \end{array}$ $\begin{array}{r} 4 \\ +8 \\ \hline \end{array}$ **M** $\begin{array}{r} 6 \\ +1 \\ \hline \end{array}$ $\begin{array}{r} 1 \\ +6 \\ \hline \end{array}$ **N** $\begin{array}{r} 5 \\ +3 \\ \hline \end{array}$ $\begin{array}{r} 3 \\ +5 \\ \hline \end{array}$

O $\begin{array}{r} 9 \\ +7 \\ \hline \end{array}$ $\begin{array}{r} 7 \\ +9 \\ \hline \end{array}$ **P** $\begin{array}{r} 5 \\ +8 \\ \hline \end{array}$ $\begin{array}{r} 8 \\ +5 \\ \hline \end{array}$ **Q** $\begin{array}{r} 6 \\ +7 \\ \hline \end{array}$ $\begin{array}{r} 7 \\ +6 \\ \hline \end{array}$

An addition fact can be turned around. The sum is the same.

$$\begin{array}{r} 8 \\ +7 \\ \hline 15 \end{array} \qquad \begin{array}{r} 7 \\ +8 \\ \hline 15 \end{array}$$

$$9 + 8 = 17$$

$$8 + 9 = 17$$

Add. Write the sum.

A $4 + 9 =$ ___ $9 + 4 =$ ___   **B** $3 + 10 =$ ___ $10 + 3 =$ ___

C $5 + 8 =$ ___ $8 + 5 =$ ___   **D** $5 + 10 =$ ___ $10 + 5 =$ ___

E $6 + 7 =$ ___ $7 + 6 =$ ___   **F** $7 + 10 =$ ___ $10 + 7 =$ ___

G $7 + 8 =$ ___ $8 + 7 =$ ___   **H** $8 + 10 =$ ___ $10 + 8 =$ ___

I $\begin{array}{r} 4 \\ +10 \end{array} \qquad \begin{array}{r} 10 \\ +4 \end{array}$ **J** $\begin{array}{r} 6 \\ +9 \end{array} \qquad \begin{array}{r} 9 \\ +6 \end{array}$

K $\begin{array}{r} 10 \\ +9 \end{array} \qquad \begin{array}{r} 9 \\ +10 \end{array}$ **L** $\begin{array}{r} 6 \\ +10 \end{array} \qquad \begin{array}{r} 10 \\ +6 \end{array}$

M $\begin{array}{r} 7 \\ +8 \end{array} \qquad \begin{array}{r} 8 \\ +7 \end{array}$ **N** $\begin{array}{r} 9 \\ +7 \end{array} \qquad \begin{array}{r} 7 \\ +9 \end{array}$

O $\begin{array}{r} 6 \\ +8 \end{array} \qquad \begin{array}{r} 8 \\ +6 \end{array}$ **P** $\begin{array}{r} 8 \\ +9 \end{array} \qquad \begin{array}{r} 9 \\ +8 \end{array}$

A
$$\begin{array}{r} 4 \\ +7 \\ \hline \end{array}$$
$$\begin{array}{r} 4 \\ +4 \\ \hline \end{array}$$
$$\begin{array}{r} 6 \\ +5 \\ \hline \end{array}$$
$$\begin{array}{r} 1 \\ +10 \\ \hline \end{array}$$
$$\begin{array}{r} 6 \\ +3 \\ \hline \end{array}$$
$$\begin{array}{r} 3 \\ +8 \\ \hline \end{array}$$

B
$$\begin{array}{r} 2 \\ +3 \\ \hline \end{array}$$
$$\begin{array}{r} 0 \\ +4 \\ \hline \end{array}$$
$$\begin{array}{r} 2 \\ +9 \\ \hline \end{array}$$
$$\begin{array}{r} 3 \\ +1 \\ \hline \end{array}$$
$$\begin{array}{r} 2 \\ +0 \\ \hline \end{array}$$
$$\begin{array}{r} 1 \\ +7 \\ \hline \end{array}$$

C
$$\begin{array}{r} 3 \\ +9 \\ \hline \end{array}$$
$$\begin{array}{r} 1 \\ +5 \\ \hline \end{array}$$
$$\begin{array}{r} 5 \\ +8 \\ \hline \end{array}$$
$$\begin{array}{r} 5 \\ +3 \\ \hline \end{array}$$
$$\begin{array}{r} 0 \\ +10 \\ \hline \end{array}$$
$$\begin{array}{r} 1 \\ +2 \\ \hline \end{array}$$

D
$$\begin{array}{r} 5 \\ +5 \\ \hline \end{array}$$
$$\begin{array}{r} 1 \\ +1 \\ \hline \end{array}$$
$$\begin{array}{r} 7 \\ +0 \\ \hline \end{array}$$
$$\begin{array}{r} 9 \\ +4 \\ \hline \end{array}$$
$$\begin{array}{r} 0 \\ +3 \\ \hline \end{array}$$
$$\begin{array}{r} 10 \\ +3 \\ \hline \end{array}$$

E
$$\begin{array}{r} 2 \\ +10 \\ \hline \end{array}$$
$$\begin{array}{r} 9 \\ +1 \\ \hline \end{array}$$
$$\begin{array}{r} 0 \\ +6 \\ \hline \end{array}$$
$$\begin{array}{r} 6 \\ +2 \\ \hline \end{array}$$
$$\begin{array}{r} 4 \\ +1 \\ \hline \end{array}$$
$$\begin{array}{r} 8 \\ +0 \\ \hline \end{array}$$

F
$$\begin{array}{r} 10 \\ +6 \\ \hline \end{array}$$
$$\begin{array}{r} 5 \\ +1 \\ \hline \end{array}$$
$$\begin{array}{r} 4 \\ +8 \\ \hline \end{array}$$
$$\begin{array}{r} 4 \\ +2 \\ \hline \end{array}$$
$$\begin{array}{r} 7 \\ +9 \\ \hline \end{array}$$
$$\begin{array}{r} 2 \\ +2 \\ \hline \end{array}$$

A

$$\begin{array}{r} 10 \\ +\ 0 \\ \hline 10 \end{array}\qquad \begin{array}{r} 4 \\ +\ 6 \\ \hline 10 \end{array}\qquad \begin{array}{r} 8 \\ +\ 2 \\ \hline 10 \end{array}\qquad \begin{array}{r} 5 \\ +10 \\ \hline 15 \end{array}\qquad \begin{array}{r} 8 \\ +\ 3 \\ \hline 11 \end{array}\qquad \begin{array}{r} 3 \\ +\ 2 \\ \hline 5 \end{array}$$

B

$$\begin{array}{r} 3 \\ +\ 3 \\ \hline 6 \end{array}\qquad \begin{array}{r} 10 \\ +\ 5 \\ \hline 15 \end{array}\qquad \begin{array}{r} 7 \\ +\ 3 \\ \hline 10 \end{array}\qquad \begin{array}{r} 7 \\ +\ 5 \\ \hline 12 \end{array}\qquad \begin{array}{r} 2 \\ +\ 6 \\ \hline 8 \end{array}\qquad \begin{array}{r} 7 \\ +\ 1 \\ \hline 8 \end{array}$$

C

$$\begin{array}{r} 1 \\ +\ 3 \\ \hline 4 \end{array}\qquad \begin{array}{r} 6 \\ +\ 4 \\ \hline 10 \end{array}\qquad \begin{array}{r} 7 \\ +\ 8 \\ \hline 15 \end{array}\qquad \begin{array}{r} 5 \\ +\ 2 \\ \hline 7 \end{array}\qquad \begin{array}{r} 10 \\ +\ 2 \\ \hline 12 \end{array}\qquad \begin{array}{r} 6 \\ +\ 8 \\ \hline 14 \end{array}$$

D

$$\begin{array}{r} 5 \\ +\ 7 \\ \hline 18 \end{array}\qquad \begin{array}{r} 4 \\ +\ 3 \\ \hline 7 \end{array}\qquad \begin{array}{r} 0 \\ +\ 1 \\ \hline 1 \end{array}\qquad \begin{array}{r} 10 \\ +\ 1 \\ \hline 11 \end{array}\qquad \begin{array}{r} 0 \\ +\ 9 \\ \hline 9 \end{array}\qquad \begin{array}{r} 6 \\ +\ 6 \\ \hline 12 \end{array}$$

E

$$\begin{array}{r} 6 \\ +\ 9 \\ \hline 15 \end{array}\qquad \begin{array}{r} 9 \\ +\ 8 \\ \hline 17 \end{array}\qquad \begin{array}{r} 5 \\ +\ 6 \\ \hline 11 \end{array}\qquad \begin{array}{r} 9 \\ +\ 7 \\ \hline 16 \end{array}\qquad \begin{array}{r} 6 \\ +\ 1 \\ \hline 7 \end{array}\qquad \begin{array}{r} 0 \\ +\ 7 \\ \hline 7 \end{array}$$

F

$$\begin{array}{r} 0 \\ +\ 8 \\ \hline 8 \end{array}\qquad \begin{array}{r} 7 \\ +\ 7 \\ \hline 14 \end{array}\qquad \begin{array}{r} 0 \\ +\ 0 \\ \hline 0 \end{array}\qquad \begin{array}{r} 10 \\ +\ 4 \\ \hline 14 \end{array}\qquad \begin{array}{r} 2 \\ +\ 5 \\ \hline 7 \end{array}\qquad \begin{array}{r} 5 \\ +\ 0 \\ \hline 5 \end{array}$$

Name_____

A
4	10	8	3	3
+5	+7	+10	+6	+7

B
7	6	0	8	1
+10	+7	+2	+7	+8

C
8	1	2	2	6
+4	+0	+7	+8	+10

D
9	10	3	8	6
+9	+10	+0	+8	+0

E
9	6	9	9	1
+3	+6	+6	+5	+9

Name_____

A
$$9 \atop +0$$ $$7 \atop +4$$ $$4 \atop +0$$ $$8 \atop +5$$ $$3 \atop +5$$

B
$$0 \atop +4$$ $$8 \atop +6$$ $$1 \atop +6$$ $$9 \atop +2$$ $$5 \atop +4$$

C
$$7 \atop +2$$ $$8 \atop +1$$ $$5 \atop +9$$ $$8 \atop +9$$ $$1 \atop +4$$

D
$$10 \atop +8$$ $$3 \atop +4$$ $$0 \atop +5$$ $$3 \atop +10$$ $$2 \atop +4$$

E
$$7 \atop +6$$ $$2 \atop +1$$ $$10 \atop +9$$ $$4 \atop +9$$ $$9 \atop +10$$

Name_____

5 − 2 = 3

6
− 1
___
5

Subtract. Write the difference.

A 2 − 1 = ___   3 − 2 = ___   1 − 0 = ___   5 − 3 = ___

B 6 − 5 = ___   1 − 1 = ___   5 − 4 = ___   4 − 2 = ___

C 3 − 0 = ___   5 − 2 = ___   6 − 3 = ___   4 − 4 = ___

D 5 − 5 = ___   6 − 4 = ___   4 − 3 = ___   3 − 1 = ___

E
$\begin{array}{r} 5 \\ -3 \\ \hline \end{array}$
$\begin{array}{r} 4 \\ -2 \\ \hline \end{array}$
$\begin{array}{r} 3 \\ -3 \\ \hline \end{array}$
$\begin{array}{r} 4 \\ -0 \\ \hline \end{array}$
$\begin{array}{r} 6 \\ -1 \\ \hline \end{array}$
$\begin{array}{r} 2 \\ -2 \\ \hline \end{array}$

F
$\begin{array}{r} 5 \\ -5 \\ \hline \end{array}$
$\begin{array}{r} 2 \\ -0 \\ \hline \end{array}$
$\begin{array}{r} 4 \\ -3 \\ \hline \end{array}$
$\begin{array}{r} 6 \\ -4 \\ \hline \end{array}$
$\begin{array}{r} 3 \\ -1 \\ \hline \end{array}$
$\begin{array}{r} 6 \\ -0 \\ \hline \end{array}$

G
$\begin{array}{r} 6 \\ -2 \\ \hline \end{array}$
$\begin{array}{r} 4 \\ -1 \\ \hline \end{array}$
$\begin{array}{r} 5 \\ -0 \\ \hline \end{array}$
$\begin{array}{r} 3 \\ -2 \\ \hline \end{array}$
$\begin{array}{r} 2 \\ -2 \\ \hline \end{array}$
$\begin{array}{r} 5 \\ -2 \\ \hline \end{array}$

H
$\begin{array}{r} 6 \\ -6 \\ \hline \end{array}$
$\begin{array}{r} 5 \\ -1 \\ \hline \end{array}$
$\begin{array}{r} 1 \\ -1 \\ \hline \end{array}$
$\begin{array}{r} 6 \\ -5 \\ \hline \end{array}$
$\begin{array}{r} 0 \\ -0 \\ \hline \end{array}$
$\begin{array}{r} 5 \\ -4 \\ \hline \end{array}$

7 − 4 = 3

8
− 2
___
6

___

Subtract. Write the difference.

A   7 − 1 = ___   8 − 3 = ___   7 − 4 = ___   8 − 0 = ___

B   8 − 4 = ___   7 − 2 = ___   8 − 1 = ___   7 − 7 = ___

C   7 − 3 = ___   8 − 7 = ___   8 − 2 = ___   8 − 5 = ___

D   7 − 5 = ___   8 − 8 = ___   7 − 6 = ___   8 − 6 = ___

E
8	7	8	7	6	8
−3	−2	−5	−0	−5	−1

F
7	8	7	8	5	7
−7	−0	−3	−4	−5	−1

G
6	8	7	8	7	5
−3	−6	−4	−2	−6	−3

H
5	8	7	6	8	6
−1	−7	−5	−4	−8	−0

Name_____

10 − 8 = 2

$$\begin{array}{r} 11 \\ -5 \\ \hline 6 \end{array}$$

Subtract. Write the difference.

A 9 − 6 = ___   11 − 3 = ___   10 − 4 = ___   12 − 8 = ___

B 12 − 4 = ___   9 − 2 = ___   11 − 1 = ___   10 − 5 = ___

C 10 − 2 = ___   12 − 6 = ___   9 − 3 = ___   11 − 7 = ___

D 12 − 9 = ___   11 − 0 = ___   9 − 9 = ___   12 − 7 = ___

E
$$\begin{array}{r} 9 \\ -8 \\ \hline \end{array} \quad \begin{array}{r} 11 \\ -10 \\ \hline \end{array} \quad \begin{array}{r} 10 \\ -9 \\ \hline \end{array} \quad \begin{array}{r} 12 \\ -5 \\ \hline \end{array} \quad \begin{array}{r} 9 \\ -2 \\ \hline \end{array} \quad \begin{array}{r} 11 \\ -4 \\ \hline \end{array}$$

F
$$\begin{array}{r} 12 \\ -7 \\ \hline \end{array} \quad \begin{array}{r} 9 \\ -1 \\ \hline \end{array} \quad \begin{array}{r} 10 \\ -3 \\ \hline \end{array} \quad \begin{array}{r} 12 \\ -2 \\ \hline \end{array} \quad \begin{array}{r} 10 \\ -10 \\ \hline \end{array} \quad \begin{array}{r} 11 \\ -6 \\ \hline \end{array}$$

G
$$\begin{array}{r} 10 \\ -7 \\ \hline \end{array} \quad \begin{array}{r} 11 \\ -8 \\ \hline \end{array} \quad \begin{array}{r} 12 \\ -3 \\ \hline \end{array} \quad \begin{array}{r} 10 \\ -0 \\ \hline \end{array} \quad \begin{array}{r} 9 \\ -4 \\ \hline \end{array} \quad \begin{array}{r} 12 \\ -10 \\ \hline \end{array}$$

H
$$\begin{array}{r} 10 \\ -6 \\ \hline \end{array} \quad \begin{array}{r} 11 \\ -9 \\ \hline \end{array} \quad \begin{array}{r} 9 \\ -5 \\ \hline \end{array} \quad \begin{array}{r} 11 \\ -2 \\ \hline \end{array} \quad \begin{array}{r} 10 \\ -1 \\ \hline \end{array} \quad \begin{array}{r} 9 \\ -7 \\ \hline \end{array}$$

14 − 6 = 8

16
−7
___
9

Subtract. Write the difference.

A 13 − 4 = __   15 − 9 = __

B 17 − 8 = ___   16 − 8 = ___

C 14 − 7 = ___   15 − 6 = ___

D 18 − 9 = ___   20 − 10 = ___

E
15	14	16	13	17
−7	−9	−7	−5	−9

F
17	13	16	14	20
−8	−6	−9	−8	−10

G
20	15	13	14	15
−20	−9	−7	−5	−8

Name_____

A number minus 0 is the number.

7 – 0 = 7

$$\begin{array}{r} 7 \\ -0 \\ \hline 7 \end{array}$$

Subtract. Write the difference.

A 9 – 0 = ___   8 – 0 = ___   5 – 5 = ___

B 1 – 0 = ___   2 – 0 = ___   3 – 3 = ___

C 2 – 2 = ___   6 – 0 = ___   9 – 0 = ___

D 5 – 0 = ___   7 – 0 = ___   1 – 1 = ___

E 4 – 0 = ___   8 – 0 = ___   9 – 9 = ___

F
$$\begin{array}{r} 7 \\ -0 \\ \hline \end{array} \qquad \begin{array}{r} 8 \\ -0 \\ \hline \end{array} \qquad \begin{array}{r} 5 \\ -5 \\ \hline \end{array} \qquad \begin{array}{r} 2 \\ -0 \\ \hline \end{array}$$

G
$$\begin{array}{r} 1 \\ -0 \\ \hline \end{array} \qquad \begin{array}{r} 5 \\ -0 \\ \hline \end{array} \qquad \begin{array}{r} 4 \\ -4 \\ \hline \end{array} \qquad \begin{array}{r} 3 \\ -0 \\ \hline \end{array}$$

H
$$\begin{array}{r} 4 \\ -0 \\ \hline \end{array} \qquad \begin{array}{r} 9 \\ -0 \\ \hline \end{array} \qquad \begin{array}{r} 10 \\ -0 \\ \hline \end{array} \qquad \begin{array}{r} 6 \\ -0 \\ \hline \end{array}$$

Name _____

A number minus the number is 0.

$5 - 5 = 0$

$$\begin{array}{r} 5 \\ -5 \\ \hline 0 \end{array}$$

Subtract. Write the difference.

A $2 - 0 =$ ____ $1 - 1 =$ ____ $3 - 3 =$ ____

B $9 - 9 =$ ____ $4 - 4 =$ ____ $2 - 2 =$ ____

C $3 - 0 =$ ____ $10 - 10 =$ ____ $7 - 7 =$ ____

D $3 - 3 =$ ____ $8 - 8 =$ ____ $7 - 0 =$ ____

E $5 - 5 =$ ____ $6 - 6 =$ ____ $4 - 0 =$ ____

F
$$\begin{array}{r} 6 \\ -6 \\ \hline \end{array} \qquad \begin{array}{r} 5 \\ -0 \\ \hline \end{array} \qquad \begin{array}{r} 7 \\ -7 \\ \hline \end{array} \qquad \begin{array}{r} 3 \\ -3 \\ \hline \end{array}$$

G
$$\begin{array}{r} 2 \\ -2 \\ \hline \end{array} \qquad \begin{array}{r} 2 \\ -2 \\ \hline \end{array} \qquad \begin{array}{r} 5 \\ -5 \\ \hline \end{array} \qquad \begin{array}{r} 4 \\ -0 \\ \hline \end{array}$$

H
$$\begin{array}{r} 9 \\ -0 \\ \hline \end{array} \qquad \begin{array}{r} 10 \\ -10 \\ \hline \end{array} \qquad \begin{array}{r} 9 \\ -9 \\ \hline \end{array} \qquad \begin{array}{r} 0 \\ -0 \\ \hline \end{array}$$

Name_____

Count back to subtract. Use a number line.

14 − 6 = 8

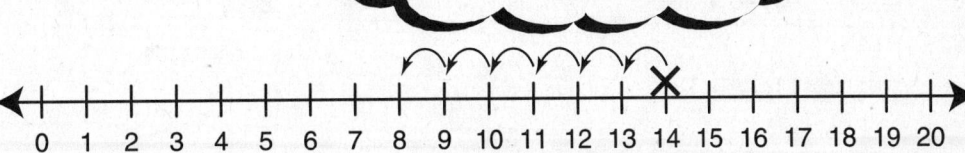

Say 14. Then count back
13, 12, 11, 10, 9, 8.

$$\begin{array}{r} 14 \\ -\ 6 \\ \hline 8 \end{array}$$

0 1 2 3 4 5 6 7 8 9 10 11 12 13 14 15 16 17 18 19 20

Use a number line to count back.
Subtract. Write the difference.

A 11 − 2 = ___   13 − 4 = ___   9 − 3 = ___

B 7 − 4 = ___   12 − 3 = ___   10 − 2 = ___

C 10 − 3 = ___   8 − 1 = ___   9 − 1 = ___

D 9 − 2 = ___   14 − 4 = ___   13 − 3 = ___

E
$$\begin{array}{r} 14 \\ -\ 4 \\ \hline \end{array}$$
$$\begin{array}{r} 9 \\ -\ 3 \\ \hline \end{array}$$
$$\begin{array}{r} 10 \\ -\ 3 \\ \hline \end{array}$$
$$\begin{array}{r} 12 \\ -\ 4 \\ \hline \end{array}$$

F
$$\begin{array}{r} 13 \\ -\ 4 \\ \hline \end{array}$$
$$\begin{array}{r} 11 \\ -\ 4 \\ \hline \end{array}$$
$$\begin{array}{r} 10 \\ -\ 4 \\ \hline \end{array}$$
$$\begin{array}{r} 8 \\ -\ 3 \\ \hline \end{array}$$

G
$$\begin{array}{r} 10 \\ -\ 1 \\ \hline \end{array}$$
$$\begin{array}{r} 9 \\ -\ 4 \\ \hline \end{array}$$
$$\begin{array}{r} 11 \\ -\ 3 \\ \hline \end{array}$$
$$\begin{array}{r} 12 \\ -\ 2 \\ \hline \end{array}$$

Related subtraction facts have the same numbers.

$$9 - 7 = 2$$
$$9 - 2 = 7$$

$$\begin{array}{r} 10 \\ -8 \\ \hline 2 \end{array}$$

$$\begin{array}{r} 10 \\ -2 \\ \hline 8 \end{array}$$

Complete the subtraction facts.

A $17 - 10 =$ ___ $17 - 7 =$ ___ **B** $12 - 7 =$ ___ $12 - 5 =$ ___

C $15 - 9 =$ ___ $15 - 6 =$ ___ **D** $9 - 6 =$ ___ $9 - 3 =$ ___

E $14 - 6 =$ ___ $14 - 8 =$ ___ **F** $5 - 3 =$ ___ $5 - 2 =$ ___

G $\begin{array}{r} 11 \\ -7 \\ \hline \end{array}$ $\begin{array}{r} 11 \\ -4 \\ \hline \end{array}$ **H** $\begin{array}{r} 9 \\ -4 \\ \hline \end{array}$ $\begin{array}{r} 9 \\ -5 \\ \hline \end{array}$ **I** $\begin{array}{r} 8 \\ -3 \\ \hline \end{array}$ $\begin{array}{r} 8 \\ -5 \\ \hline \end{array}$

J $\begin{array}{r} 13 \\ -8 \\ \hline \end{array}$ $\begin{array}{r} 13 \\ -5 \\ \hline \end{array}$ **K** $\begin{array}{r} 8 \\ -2 \\ \hline \end{array}$ $\begin{array}{r} 8 \\ -6 \\ \hline \end{array}$ **L** $\begin{array}{r} 12 \\ -8 \\ \hline \end{array}$ $\begin{array}{r} 12 \\ -4 \\ \hline \end{array}$

M $\begin{array}{r} 19 \\ -9 \\ \hline \end{array}$ $\begin{array}{r} 19 \\ -10 \\ \hline \end{array}$ **N** $\begin{array}{r} 10 \\ -6 \\ \hline \end{array}$ $\begin{array}{r} 10 \\ -4 \\ \hline \end{array}$ **O** $\begin{array}{r} 16 \\ -9 \\ \hline \end{array}$ $\begin{array}{r} 16 \\ -7 \\ \hline \end{array}$

Use addition near doubles to subtract.

$$
\begin{array}{r} 5 \\ +5 \\ \hline 10 \end{array}
\qquad
\begin{array}{r} 10 \\ -5 \\ \hline 5 \end{array}
\qquad
\begin{array}{r} 10 \\ -6 \\ \hline 4 \end{array}
\qquad
\begin{array}{r} 10 \\ -4 \\ \hline 6 \end{array}
$$

Write the sum or the difference.

A $7 + 7 = $ ___ $14 - 6 = $ ___ $14 - 8 = $ ___

B $4 + 4 = $ ___ $8 - 3 = $ ___ $8 - 5 = $ ___

C $9 + 9 = $ ___ $18 - 10 = $ ___ $18 - 8 = $ ___

D $3 + 3 = $ ___ $6 - 4 = $ ___ $6 - 2 = $ ___

E
$$
\begin{array}{r} 20 \\ -10 \\ \hline \end{array}
\qquad
\begin{array}{r} 20 \\ -11 \\ \hline \end{array}
\qquad
\begin{array}{r} 20 \\ -9 \\ \hline \end{array}
$$
F
$$
\begin{array}{r} 16 \\ -8 \\ \hline \end{array}
\qquad
\begin{array}{r} 16 \\ -9 \\ \hline \end{array}
\qquad
\begin{array}{r} 16 \\ -7 \\ \hline \end{array}
$$

G
$$
\begin{array}{r} 4 \\ -2 \\ \hline \end{array}
\qquad
\begin{array}{r} 4 \\ -1 \\ \hline \end{array}
\qquad
\begin{array}{r} 4 \\ -3 \\ \hline \end{array}
$$
H
$$
\begin{array}{r} 14 \\ -7 \\ \hline \end{array}
\qquad
\begin{array}{r} 14 \\ -8 \\ \hline \end{array}
\qquad
\begin{array}{r} 14 \\ -6 \\ \hline \end{array}
$$

I
$$
\begin{array}{r} 18 \\ -9 \\ \hline \end{array}
\qquad
\begin{array}{r} 18 \\ -10 \\ \hline \end{array}
\qquad
\begin{array}{r} 18 \\ -8 \\ \hline \end{array}
$$
J
$$
\begin{array}{r} 12 \\ -6 \\ \hline \end{array}
\qquad
\begin{array}{r} 12 \\ -5 \\ \hline \end{array}
\qquad
\begin{array}{r} 12 \\ -7 \\ \hline \end{array}
$$

Make a ten. Then subtract the rest of the second number.

$$\begin{array}{r} 13 \\ -5 \\ \hline 8 \end{array}$$

$$13 - 5 = 8$$

Say $13 - 3 = 10$ and $5 - 3 = 2$
That means that $13 - 5$ equals $10 - 2$.
$10 - 2 = 8$, so $13 - 5 = 8$.

Make a ten. Then find the difference.

A $11 - 8 =$ ___ $14 - 9 =$ ___ $18 - 9 =$ ___ $14 - 6 =$ ___

B $15 - 7 =$ ___ $13 - 6 =$ ___ $11 - 3 =$ ___ $12 - 2 =$ ___

C $12 - 4 =$ ___ $15 - 8 =$ ___ $16 - 7 =$ ___ $11 - 5 =$ ___

D
$$\begin{array}{r} 11 \\ -2 \\ \hline \end{array} \qquad \begin{array}{r} 16 \\ -9 \\ \hline \end{array} \qquad \begin{array}{r} 12 \\ -5 \\ \hline \end{array} \qquad \begin{array}{r} 11 \\ -6 \\ \hline \end{array} \qquad \begin{array}{r} 17 \\ -8 \\ \hline \end{array}$$

E
$$\begin{array}{r} 11 \\ -7 \\ \hline \end{array} \qquad \begin{array}{r} 13 \\ -7 \\ \hline \end{array} \qquad \begin{array}{r} 17 \\ -8 \\ \hline \end{array} \qquad \begin{array}{r} 15 \\ -6 \\ \hline \end{array} \qquad \begin{array}{r} 12 \\ -3 \\ \hline \end{array}$$

F
$$\begin{array}{r} 13 \\ -8 \\ \hline \end{array} \qquad \begin{array}{r} 14 \\ -5 \\ \hline \end{array} \qquad \begin{array}{r} 13 \\ -5 \\ \hline \end{array} \qquad \begin{array}{r} 14 \\ -8 \\ \hline \end{array} \qquad \begin{array}{r} 11 \\ -9 \\ \hline \end{array}$$

G
$$\begin{array}{r} 13 \\ -4 \\ \hline \end{array} \qquad \begin{array}{r} 12 \\ -8 \\ \hline \end{array} \qquad \begin{array}{r} 11 \\ -4 \\ \hline \end{array} \qquad \begin{array}{r} 12 \\ -7 \\ \hline \end{array} \qquad \begin{array}{r} 15 \\ -9 \\ \hline \end{array}$$

Name_____

The difference may be greater than the number being subtracted.

$$8 - 2 = 6$$

$$\begin{array}{r} 8 \\ -2 \\ \hline 6 \end{array}$$

Subtract. Write the difference.

A $9 - 1 =$ ___ $12 - 3 =$ ___ $19 - 9 =$ ___ $12 - 5 =$ ___

B $11 - 5 =$ ___ $14 - 4 =$ ___ $10 - 2 =$ ___ $6 - 2 =$ ___

C $9 - 2 =$ ___ $13 - 5 =$ ___ $11 - 3 =$ ___ $15 - 6 =$ ___

D $7 - 2 =$ ___ $8 - 3 =$ ___ $4 - 1 =$ ___ $16 - 7 =$ ___

E
$$\begin{array}{r} 11 \\ -3 \\ \hline \end{array} \qquad \begin{array}{r} 3 \\ -0 \\ \hline \end{array} \qquad \begin{array}{r} 17 \\ -8 \\ \hline \end{array} \qquad \begin{array}{r} 13 \\ -6 \\ \hline \end{array} \qquad \begin{array}{r} 10 \\ -3 \\ \hline \end{array} \qquad \begin{array}{r} 9 \\ -0 \\ \hline \end{array}$$

F
$$\begin{array}{r} 7 \\ -1 \\ \hline \end{array} \qquad \begin{array}{r} 14 \\ -5 \\ \hline \end{array} \qquad \begin{array}{r} 5 \\ -2 \\ \hline \end{array} \qquad \begin{array}{r} 9 \\ -4 \\ \hline \end{array} \qquad \begin{array}{r} 6 \\ -1 \\ \hline \end{array} \qquad \begin{array}{r} 10 \\ -2 \\ \hline \end{array}$$

G
$$\begin{array}{r} 10 \\ -4 \\ \hline \end{array} \qquad \begin{array}{r} 8 \\ -2 \\ \hline \end{array} \qquad \begin{array}{r} 12 \\ -4 \\ \hline \end{array} \qquad \begin{array}{r} 18 \\ -8 \\ \hline \end{array} \qquad \begin{array}{r} 13 \\ -4 \\ \hline \end{array} \qquad \begin{array}{r} 8 \\ -3 \\ \hline \end{array}$$

The difference may be the same as the number being subtracted.

$$\begin{array}{r} 14 \\ -7 \\ \hline 7 \end{array}$$

Subtract. Write the difference.

A $10 - 5 = \underline{}$ $\qquad$ $4 - 2 = \underline{}$

B $18 - 9 = \underline{}$ $\qquad$ $12 - 6 = \underline{}$

C $8 - 4 = \underline{}$ $\qquad$ $6 - 3 = \underline{}$

D $2 - 1 = \underline{}$ $\qquad$ $0 - 0 = \underline{}$

E
$$\begin{array}{r} 14 \\ -7 \\ \hline \end{array} \qquad \begin{array}{r} 0 \\ -0 \\ \hline \end{array} \qquad \begin{array}{r} 6 \\ -3 \\ \hline \end{array} \qquad \begin{array}{r} 2 \\ -1 \\ \hline \end{array}$$

F
$$\begin{array}{r} 18 \\ -9 \\ \hline \end{array} \qquad \begin{array}{r} 6 \\ -3 \\ \hline \end{array} \qquad \begin{array}{r} 4 \\ -2 \\ \hline \end{array} \qquad \begin{array}{r} 12 \\ -6 \\ \hline \end{array}$$

G
$$\begin{array}{r} 10 \\ -5 \\ \hline \end{array} \qquad \begin{array}{r} 16 \\ -8 \\ \hline \end{array} \qquad \begin{array}{r} 8 \\ -4 \\ \hline \end{array} \qquad \begin{array}{r} 20 \\ -10 \\ \hline \end{array}$$

Name_____

The difference may be less than the number being subtracted.

7 − 5 = 2

$$\begin{array}{r} 7 \\ -5 \\ \hline 2 \end{array}$$

Subtract. Write the difference.

A 12 − 9 = ___   11 − 7 = ___   7 − 5 = ___   10 − 7 = ___

B 10 − 9 = ___   7 − 4 = ___   4 − 4 = ___   12 − 8 = ___

C 9 − 7 = ___   12 − 7 = ___   6 − 5 = ___   8 − 6 = ___

D 4 − 3 = ___   9 − 8 = ___   15 − 10 = ___   17 − 9 = ___

E
$$\begin{array}{r} 15 \\ -8 \\ \hline \end{array} \qquad \begin{array}{r} 12 \\ -7 \\ \hline \end{array} \qquad \begin{array}{r} 5 \\ -3 \\ \hline \end{array} \qquad \begin{array}{r} 7 \\ -6 \\ \hline \end{array} \qquad \begin{array}{r} 9 \\ -8 \\ \hline \end{array} \qquad \begin{array}{r} 13 \\ -9 \\ \hline \end{array}$$

F
$$\begin{array}{r} 5 \\ -4 \\ \hline \end{array} \qquad \begin{array}{r} 3 \\ -2 \\ \hline \end{array} \qquad \begin{array}{r} 8 \\ -5 \\ \hline \end{array} \qquad \begin{array}{r} 13 \\ -7 \\ \hline \end{array} \qquad \begin{array}{r} 10 \\ -6 \\ \hline \end{array} \qquad \begin{array}{r} 11 \\ -9 \\ \hline \end{array}$$

G
$$\begin{array}{r} 11 \\ -8 \\ \hline \end{array} \qquad \begin{array}{r} 14 \\ -9 \\ \hline \end{array} \qquad \begin{array}{r} 9 \\ -6 \\ \hline \end{array} \qquad \begin{array}{r} 17 \\ -9 \\ \hline \end{array} \qquad \begin{array}{r} 6 \\ -4 \\ \hline \end{array} \qquad \begin{array}{r} 13 \\ -7 \\ \hline \end{array}$$

A
4	7	16	9	8
−1	−5	−7	−2	−7

B
12	8	14	12	15
−0	−5	−6	−9	−10

C
3	17	10	13	4
−1	−7	−0	−5	−3

D
6	13	8	11	11
−6	−7	−0	−10	−5

E
8	18	9	10	14
−3	−8	−4	−6	−10

F
9	6	15	7	13
−0	−2	−9	−4	−10

A
$$\begin{array}{r} 13 \\ -\ 4 \\ \hline \end{array}$$
$$\begin{array}{r} 16 \\ -10 \\ \hline \end{array}$$
$$\begin{array}{r} 7 \\ -1 \\ \hline \end{array}$$
$$\begin{array}{r} 12 \\ -\ 7 \\ \hline \end{array}$$
$$\begin{array}{r} 20 \\ -10 \\ \hline \end{array}$$

B
$$\begin{array}{r} 12 \\ -\ 1 \\ \hline \end{array}$$
$$\begin{array}{r} 14 \\ -\ 7 \\ \hline \end{array}$$
$$\begin{array}{r} 16 \\ -\ 9 \\ \hline \end{array}$$
$$\begin{array}{r} 16 \\ -\ 6 \\ \hline \end{array}$$
$$\begin{array}{r} 18 \\ -\ 9 \\ \hline \end{array}$$

C
$$\begin{array}{r} 10 \\ -\ 7 \\ \hline \end{array}$$
$$\begin{array}{r} 11 \\ -\ 0 \\ \hline \end{array}$$
$$\begin{array}{r} 13 \\ -\ 9 \\ \hline \end{array}$$
$$\begin{array}{r} 14 \\ -\ 5 \\ \hline \end{array}$$
$$\begin{array}{r} 10 \\ -\ 1 \\ \hline \end{array}$$

D
$$\begin{array}{r} 15 \\ -\ 5 \\ \hline \end{array}$$
$$\begin{array}{r} 11 \\ -\ 9 \\ \hline \end{array}$$
$$\begin{array}{r} 11 \\ -\ 1 \\ \hline \end{array}$$
$$\begin{array}{r} 12 \\ -\ 2 \\ \hline \end{array}$$
$$\begin{array}{r} 11 \\ -\ 6 \\ \hline \end{array}$$

E
$$\begin{array}{r} 11 \\ -\ 7 \\ \hline \end{array}$$
$$\begin{array}{r} 13 \\ -\ 6 \\ \hline \end{array}$$
$$\begin{array}{r} 10 \\ -\ 5 \\ \hline \end{array}$$
$$\begin{array}{r} 15 \\ -\ 9 \\ \hline \end{array}$$
$$\begin{array}{r} 10 \\ -\ 3 \\ \hline \end{array}$$

Name _____

**Power Drill
Subtraction Facts**

A
$$\begin{array}{r} 2 \\ -0 \\ \hline \end{array} \qquad \begin{array}{r} 5 \\ -3 \\ \hline \end{array} \qquad \begin{array}{r} 8 \\ -1 \\ \hline \end{array} \qquad \begin{array}{r} 15 \\ -8 \\ \hline \end{array} \qquad \begin{array}{r} 15 \\ -6 \\ \hline \end{array} \qquad \begin{array}{r} 7 \\ -7 \\ \hline \end{array}$$

B
$$\begin{array}{r} 6 \\ -5 \\ \hline \end{array} \qquad \begin{array}{r} 9 \\ -1 \\ \hline \end{array} \qquad \begin{array}{r} 7 \\ -3 \\ \hline \end{array} \qquad \begin{array}{r} 12 \\ -4 \\ \hline \end{array} \qquad \begin{array}{r} 12 \\ -8 \\ \hline \end{array} \qquad \begin{array}{r} 16 \\ -8 \\ \hline \end{array}$$

C
$$\begin{array}{r} 19 \\ -10 \\ \hline \end{array} \qquad \begin{array}{r} 8 \\ -8 \\ \hline \end{array} \qquad \begin{array}{r} 3 \\ -3 \\ \hline \end{array} \qquad \begin{array}{r} 19 \\ -9 \\ \hline \end{array} \qquad \begin{array}{r} 13 \\ -3 \\ \hline \end{array} \qquad \begin{array}{r} 12 \\ -5 \\ \hline \end{array}$$

D
$$\begin{array}{r} 17 \\ -8 \\ \hline \end{array} \qquad \begin{array}{r} 6 \\ -2 \\ \hline \end{array} \qquad \begin{array}{r} 6 \\ -0 \\ \hline \end{array} \qquad \begin{array}{r} 14 \\ -4 \\ \hline \end{array} \qquad \begin{array}{r} 14 \\ -9 \\ \hline \end{array} \qquad \begin{array}{r} 5 \\ -0 \\ \hline \end{array}$$

E
$$\begin{array}{r} 9 \\ -6 \\ \hline \end{array} \qquad \begin{array}{r} 6 \\ -4 \\ \hline \end{array} \qquad \begin{array}{r} 17 \\ -10 \\ \hline \end{array} \qquad \begin{array}{r} 10 \\ -9 \\ \hline \end{array} \qquad \begin{array}{r} 18 \\ -10 \\ \hline \end{array} \qquad \begin{array}{r} 9 \\ -5 \\ \hline \end{array}$$

F
$$\begin{array}{r} 10 \\ -8 \\ \hline \end{array} \qquad \begin{array}{r} 2 \\ -2 \\ \hline \end{array} \qquad \begin{array}{r} 1 \\ -1 \\ \hline \end{array} \qquad \begin{array}{r} 4 \\ -0 \\ \hline \end{array} \qquad \begin{array}{r} 17 \\ -9 \\ \hline \end{array} \qquad \begin{array}{r} 11 \\ -4 \\ \hline \end{array}$$

© McGraw-Hill School Division

Name _____

A

4	10	5	11	15	11
−4	−10	−5	−2	−7	−8

B

6	3	7	10	14	1
−3	−0	−4	−4	−8	−0

C

8	5	8	7	5	9
−4	−4	−6	−2	−2	−7

D

2	7	9	12	9	10
−1	−6	−3	−6	−9	−2

E

3	9	6	11	13	0
−2	−8	−1	−3	−8	−0

F

7	4	12	8	12	5
−0	−2	−10	−2	−3	−1

Name_____

Use fact families to add and subtract.

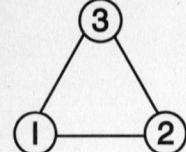

$1 + 2 = 3$ $3 - 2 = 1$
$2 + 1 = 3$ $3 - 1 = 1$

Add or subtract.
Complete the facts that make each family.

A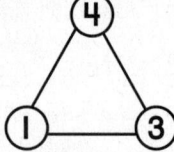

$1 + 3 = $ ___ $4 - 3 = $ ___

$3 + 1 = $ ___ $4 - 1 = $ ___

B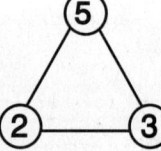

$2 + 3 = $ ___ $5 - 3 = $ ___

$3 + 2 = $ ___ $5 - 2 = $ ___

C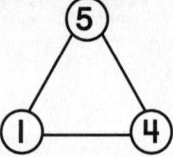

$1 + 4 = $ ___ $5 - 4 = $ ___

$4 + 1 = $ ___ $5 - 1 = $ ___

D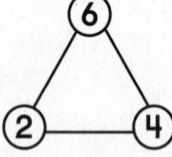

$2 + 4 = $ ___ $6 - 4 = $ ___

$4 + 2 = $ ___ $6 - 2 = $ ___

E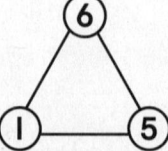

$1 + 5 = $ ___ $6 - 5 = $ ___

$5 + 1 = $ ___ $6 - 1 = $ ___

Name_____

Use fact families to add and subtract.

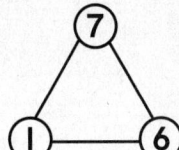

$$1 + 6 = 7 \qquad 7 - 6 = 1$$
$$6 + 1 = 7 \qquad 7 - 3 = 6$$

Add or subtract.
Complete the facts that make each family.

A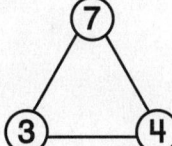

$$3 + 4 = \underline{\hspace{1cm}} \qquad 7 - 4 = \underline{\hspace{1cm}}$$
$$4 + 3 = \underline{\hspace{1cm}} \qquad 7 - 3 = \underline{\hspace{1cm}}$$

B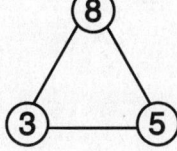

$$3 + 5 = \underline{\hspace{1cm}} \qquad 8 - 5 = \underline{\hspace{1cm}}$$
$$5 + 3 = \underline{\hspace{1cm}} \qquad 8 - 3 = \underline{\hspace{1cm}}$$

C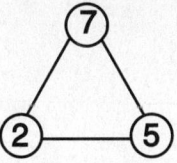

$$2 + 5 = \underline{\hspace{1cm}} \qquad 7 - 5 = \underline{\hspace{1cm}}$$
$$5 + 2 = \underline{\hspace{1cm}} \qquad 7 - 2 = \underline{\hspace{1cm}}$$

D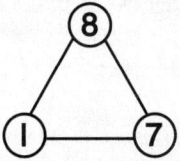

$$1 + 7 = \underline{\hspace{1cm}} \qquad 8 - 7 = \underline{\hspace{1cm}}$$
$$7 + 1 = \underline{\hspace{1cm}} \qquad 8 - 1 = \underline{\hspace{1cm}}$$

E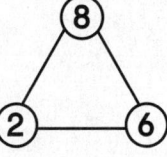

$$2 + 6 = \underline{\hspace{1cm}} \qquad 8 - 6 = \underline{\hspace{1cm}}$$
$$6 + 2 = \underline{\hspace{1cm}} \qquad 8 - 2 = \underline{\hspace{1cm}}$$

Name_____

Use fact families to add and subtract.

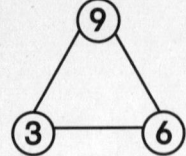

$3 + 6 = 9$ $9 - 6 = 3$
$6 + 3 = 9$ $9 - 3 = 6$

Add or subtract.

Complete the facts that make each family.

A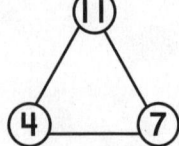

$4 + 7 = $ ___ $11 - 7 = $ ___

$7 + 4 = $ ___ $11 - 4 = $ ___

B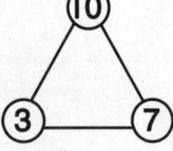

$3 + 7 = $ ___ $10 - 7 = $ ___

$7 + 3 = $ ___ $10 - 3 = $ ___

C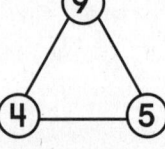

$4 + 5 = $ ___ $9 - 5 = $ ___

$5 + 4 = $ ___ $9 - 4 = $ ___

D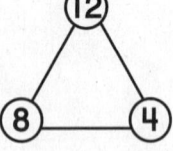

$4 + 8 = $ ___ $12 - 8 = $ ___

$8 + 4 = $ ___ $12 - 4 = $ ___

E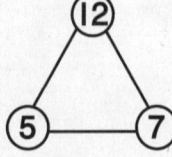

$5 + 7 = $ ___ $12 - 5 = $ ___

$7 + 5 = $ ___ $12 - 7 = $ ___

Name_____

Use fact families to add and subtract.

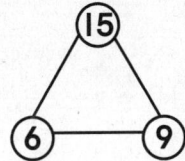

$9 + 6 = 15$	$15 - 6 = 9$
$6 + 9 = 15$	$15 - 9 = 6$

Add or subtract.

Complete the facts that make each family.

A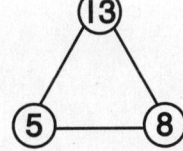

$5 + 8 =$ ___ $13 - 5 =$ ___

$8 + 5 =$ ___ $13 - 8 =$ ___

B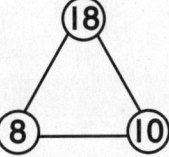

$8 + 10 =$ ___ $18 - 10 =$ ___

$10 + 8 =$ ___ $18 - 8 =$ ___

C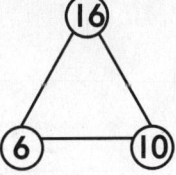

$6 + 10 =$ ___ $16 - 10 =$ ___

$10 + 6 =$ ___ $16 - 6 =$ ___

D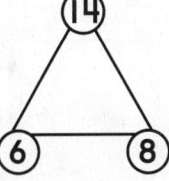

$6 + 8 =$ ___ $14 - 6 =$ ___

$8 + 6 =$ ___ $14 - 8 =$ ___

E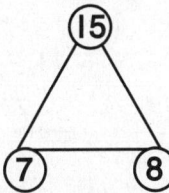

$7 + 8 =$ ___ $15 - 8 =$ ___

$8 + 7 =$ ___ $15 - 7 =$ ___

Name _____

Add or subtract. Write the sum or difference.

A $5 + 10 = $ ___ $2 + 0 = $ ___ $3 + 9 = $ ___ $6 + 2 = $ ___

B $9 - 2 = $ ___ $11 - 5 = $ ___ $2 - 1 = $ ___ $15 - 10 = $ ___

C $5 + 3 = $ ___ $9 + 1 = $ ___ $0 + 6 = $ ___ $2 + 8 = $ ___

D $13 - 5 = $ ___ $11 - 10 = $ ___ $10 - 6 = $ ___ $12 - 9 = $ ___

E $7 + 9 = $ ___ $10 + 10 = $ ___ $4 + 8 = $ ___ $0 + 0 = $ ___

F
$$\begin{array}{r} 8 \\ -5 \\ \hline \end{array} \qquad \begin{array}{r} 8 \\ -7 \\ \hline \end{array} \qquad \begin{array}{r} 7 \\ -5 \\ \hline \end{array} \qquad \begin{array}{r} 10 \\ -0 \\ \hline \end{array}$$

G
$$\begin{array}{r} 5 \\ +5 \\ \hline \end{array} \qquad \begin{array}{r} 3 \\ +1 \\ \hline \end{array} \qquad \begin{array}{r} 8 \\ +8 \\ \hline \end{array} \qquad \begin{array}{r} 2 \\ +2 \\ \hline \end{array}$$

H
$$\begin{array}{r} 9 \\ -6 \\ \hline \end{array} \qquad \begin{array}{r} 12 \\ -2 \\ \hline \end{array} \qquad \begin{array}{r} 14 \\ -6 \\ \hline \end{array} \qquad \begin{array}{r} 8 \\ -0 \\ \hline \end{array}$$

I
$$\begin{array}{r} 9 \\ +8 \\ \hline \end{array} \qquad \begin{array}{r} 10 \\ +0 \\ \hline \end{array} \qquad \begin{array}{r} 10 \\ +2 \\ \hline \end{array} \qquad \begin{array}{r} 4 \\ +4 \\ \hline \end{array}$$

J
$$\begin{array}{r} 19 \\ -9 \\ \hline \end{array} \qquad \begin{array}{r} 18 \\ -8 \\ \hline \end{array} \qquad \begin{array}{r} 3 \\ -1 \\ \hline \end{array} \qquad \begin{array}{r} 13 \\ -7 \\ \hline \end{array}$$

Name _____

Add or subtract. Write the sum or difference.

A $7 + 2 =$ ___ $8 + 7 =$ ___ $3 + 0 =$ ___ $9 + 9 =$ ___

B $6 - 2 =$ ___ $6 - 0 =$ ___ $8 - 3 =$ ___ $7 - 7 =$ ___

C $10 + 3 =$ ___ $8 + 4 =$ ___ $6 + 1 =$ ___ $7 + 8 =$ ___

D $1 - 1 =$ ___ $17 - 10 =$ ___ $7 - 3 =$ ___ $3 - 3 =$ ___

E $10 - 8 =$ ___ $5 - 3 =$ ___ $14 - 4 =$ ___ $15 - 8 =$ ___

F
$$\begin{array}{r} 4 \\ +6 \\ \hline \end{array} \qquad \begin{array}{r} 10 \\ +6 \\ \hline \end{array} \qquad \begin{array}{r} 8 \\ +0 \\ \hline \end{array} \qquad \begin{array}{r} 3 \\ +7 \\ \hline \end{array}$$

G
$$\begin{array}{r} 6 \\ -6 \\ \hline \end{array} \qquad \begin{array}{r} 4 \\ -1 \\ \hline \end{array} \qquad \begin{array}{r} 5 \\ -0 \\ \hline \end{array} \qquad \begin{array}{r} 9 \\ -0 \\ \hline \end{array}$$

H
$$\begin{array}{r} 5 \\ +4 \\ \hline \end{array} \qquad \begin{array}{r} 6 \\ +0 \\ \hline \end{array} \qquad \begin{array}{r} 7 \\ +7 \\ \hline \end{array} \qquad \begin{array}{r} 4 \\ +5 \\ \hline \end{array}$$

I
$$\begin{array}{r} 15 \\ -6 \\ \hline \end{array} \qquad \begin{array}{r} 2 \\ -0 \\ \hline \end{array} \qquad \begin{array}{r} 14 \\ -9 \\ \hline \end{array} \qquad \begin{array}{r} 8 \\ -1 \\ \hline \end{array}$$

J
$$\begin{array}{r} 4 \\ +1 \\ \hline \end{array} \qquad \begin{array}{r} 7 \\ +6 \\ \hline \end{array} \qquad \begin{array}{r} 9 \\ +2 \\ \hline \end{array} \qquad \begin{array}{r} 5 \\ +9 \\ \hline \end{array}$$

Add or subtract. Write the sum or difference.

A $5 + 8 =$ ___ $10 + 9 =$ ___ $4 + 2 =$ ___ $0 + 5 =$ ___

B $7 - 0 =$ ___ $5 - 2 =$ ___ $1 - 0 =$ ___ $12 - 3 =$ ___

C $9 + 7 =$ ___ $0 + 10 =$ ___ $6 + 7 =$ ___ $9 + 0 =$ ___

D $11 - 8 =$ ___ $13 - 8 =$ ___ $9 - 9 =$ ___ $8 - 4 =$ ___

E $0 + 3 =$ ___ $8 + 9 =$ ___ $1 + 10 =$ ___ $2 + 4 =$ ___

F
$$\begin{array}{r} 10 \\ -4 \\ \hline \end{array} \qquad \begin{array}{r} 7 \\ -4 \\ \hline \end{array} \qquad \begin{array}{r} 13 \\ -4 \\ \hline \end{array} \qquad \begin{array}{r} 16 \\ -7 \\ \hline \end{array}$$

G
$$\begin{array}{r} 2 \\ +5 \\ \hline \end{array} \qquad \begin{array}{r} 3 \\ +2 \\ \hline \end{array} \qquad \begin{array}{r} 9 \\ +10 \\ \hline \end{array} \qquad \begin{array}{r} 9 \\ +3 \\ \hline \end{array}$$

H
$$\begin{array}{r} 10 \\ -7 \\ \hline \end{array} \qquad \begin{array}{r} 14 \\ -8 \\ \hline \end{array} \qquad \begin{array}{r} 11 \\ -7 \\ \hline \end{array} \qquad \begin{array}{r} 15 \\ -7 \\ \hline \end{array}$$

I
$$\begin{array}{r} 2 \\ +6 \\ \hline \end{array} \qquad \begin{array}{r} 0 \\ +4 \\ \hline \end{array} \qquad \begin{array}{r} 3 \\ +5 \\ \hline \end{array} \qquad \begin{array}{r} 8 \\ +1 \\ \hline \end{array}$$

J
$$\begin{array}{r} 11 \\ -2 \\ \hline \end{array} \qquad \begin{array}{r} 3 \\ -0 \\ \hline \end{array} \qquad \begin{array}{r} 5 \\ -5 \\ \hline \end{array} \qquad \begin{array}{r} 10 \\ -10 \\ \hline \end{array}$$

Name_____

Add or subtract. Write the sum or difference.

A $6 - 3 =$ ___ $3 - 2 =$ ___ $5 - 1 =$ ___ $4 - 4 =$ ___

B $4 + 7 =$ ___ $0 + 2 =$ ___ $4 + 9 =$ ___ $7 + 4 =$ ___

C $9 - 7 =$ ___ $5 - 4 =$ ___ $12 - 10 =$ ___ $9 - 8 =$ ___

D $5 + 5 =$ ___ $3 + 10 =$ ___ $6 + 6 =$ ___ $1 + 0 =$ ___

E $10 - 2 =$ ___ $6 - 1 =$ ___ $9 - 3 =$ ___ $8 - 6 =$ ___

F
$$\begin{array}{r} 9 \\ +6 \\ \hline \end{array} \qquad \begin{array}{r} 5 \\ +0 \\ \hline \end{array} \qquad \begin{array}{r} 1 \\ +9 \\ \hline \end{array} \qquad \begin{array}{r} 4 \\ +3 \\ \hline \end{array}$$

G
$$\begin{array}{r} 4 \\ -2 \\ \hline \end{array} \qquad \begin{array}{r} 12 \\ -6 \\ \hline \end{array} \qquad \begin{array}{r} 7 \\ -2 \\ \hline \end{array} \qquad \begin{array}{r} 11 \\ -3 \\ \hline \end{array}$$

H
$$\begin{array}{r} 2 \\ +3 \\ \hline \end{array} \qquad \begin{array}{r} 4 \\ +9 \\ \hline \end{array} \qquad \begin{array}{r} 8 \\ +5 \\ \hline \end{array} \qquad \begin{array}{r} 3 \\ +10 \\ \hline \end{array}$$

I
$$\begin{array}{r} 8 \\ -2 \\ \hline \end{array} \qquad \begin{array}{r} 7 \\ -6 \\ \hline \end{array} \qquad \begin{array}{r} 0 \\ -0 \\ \hline \end{array} \qquad \begin{array}{r} 14 \\ -10 \\ \hline \end{array}$$

J
$$\begin{array}{r} 1 \\ +8 \\ \hline \end{array} \qquad \begin{array}{r} 6 \\ +3 \\ \hline \end{array} \qquad \begin{array}{r} 10 \\ +7 \\ \hline \end{array} \qquad \begin{array}{r} 1 \\ +1 \\ \hline \end{array}$$

Add or subtract. Write the sum or difference.

A $11 - 9 =$ ___ $13 - 6 =$ ___ $14 - 5 =$ ___ $17 - 7 =$ ___

B $1 + 4 =$ ___ $8 + 6 =$ ___ $5 + 2 =$ ___ $6 + 10 =$ ___

C $10 - 5 =$ ___ $11 - 1 =$ ___ $11 - 4 =$ ___ $16 - 9 =$ ___

D $2 + 7 =$ ___ $7 + 10 =$ ___ $1 + 5 =$ ___ $3 + 6 =$ ___

E $15 - 5 =$ ___ $9 - 5 =$ ___ $16 - 8 =$ ___ $12 - 5 =$ ___

F
$$\begin{array}{r} 12 \\ -8 \\ \hline \end{array} \qquad \begin{array}{r} 18 \\ -10 \\ \hline \end{array} \qquad \begin{array}{r} 17 \\ -9 \\ \hline \end{array} \qquad \begin{array}{r} 13 \\ -3 \\ \hline \end{array}$$

G
$$\begin{array}{r} 10 \\ +4 \\ \hline \end{array} \qquad \begin{array}{r} 0 \\ +8 \\ \hline \end{array} \qquad \begin{array}{r} 7 \\ +1 \\ \hline \end{array} \qquad \begin{array}{r} 8 \\ +10 \\ \hline \end{array}$$

H
$$\begin{array}{r} 4 \\ -0 \\ \hline \end{array} \qquad \begin{array}{r} 10 \\ -9 \\ \hline \end{array} \qquad \begin{array}{r} 12 \\ -4 \\ \hline \end{array} \qquad \begin{array}{r} 13 \\ -9 \\ \hline \end{array}$$

I
$$\begin{array}{r} 2 \\ +9 \\ \hline \end{array} \qquad \begin{array}{r} 7 \\ +0 \\ \hline \end{array} \qquad \begin{array}{r} 10 \\ +5 \\ \hline \end{array} \qquad \begin{array}{r} 1 \\ +2 \\ \hline \end{array}$$

J
$$\begin{array}{r} 6 \\ -4 \\ \hline \end{array} \qquad \begin{array}{r} 2 \\ -2 \\ \hline \end{array} \qquad \begin{array}{r} 9 \\ -1 \\ \hline \end{array} \qquad \begin{array}{r} 8 \\ -8 \\ \hline \end{array}$$

Name _____

Add or subtract. Write the sum or difference.

A $5 + 6 =$ ___ $9 + 4 =$ ___ $1 + 6 =$ ___ $7 + 5 =$ ___

B $6 - 5 =$ ___ $17 - 8 =$ ___ $9 - 6 =$ ___ $19 - 10 =$ ___

C $9 + 5 =$ ___ $1 + 3 =$ ___ $2 + 1 =$ ___ $3 + 8 =$ ___

D $8 + 3 =$ ___ $3 + 3 =$ ___ $6 + 8 =$ ___ $4 + 0 =$ ___

E $6 + 4 =$ ___ $0 + 7 =$ ___ $10 + 8 =$ ___ $0 + 1 =$ ___

F
$$\begin{array}{r} 11 \\ -6 \\ \hline \end{array} \qquad \begin{array}{r} 13 \\ -10 \\ \hline \end{array} \qquad \begin{array}{r} 10 \\ -3 \\ \hline \end{array} \qquad \begin{array}{r} 20 \\ -10 \\ \hline \end{array} \qquad \begin{array}{r} 18 \\ -9 \\ \hline \end{array}$$

G
$$\begin{array}{r} 8 \\ +2 \\ \hline \end{array} \qquad \begin{array}{r} 5 \\ +7 \\ \hline \end{array} \qquad \begin{array}{r} 0 \\ +0 \\ \hline \end{array} \qquad \begin{array}{r} 4 \\ +10 \\ \hline \end{array} \qquad \begin{array}{r} 10 \\ +1 \\ \hline \end{array}$$

H
$$\begin{array}{r} 12 \\ -7 \\ \hline \end{array} \qquad \begin{array}{r} 14 \\ -7 \\ \hline \end{array} \qquad \begin{array}{r} 16 \\ -10 \\ \hline \end{array} \qquad \begin{array}{r} 15 \\ -9 \\ \hline \end{array} \qquad \begin{array}{r} 16 \\ -6 \\ \hline \end{array}$$

I
$$\begin{array}{r} 0 \\ +9 \\ \hline \end{array} \qquad \begin{array}{r} 3 \\ +4 \\ \hline \end{array} \qquad \begin{array}{r} 6 \\ +5 \\ \hline \end{array} \qquad \begin{array}{r} 6 \\ +9 \\ \hline \end{array} \qquad \begin{array}{r} 7 \\ +3 \\ \hline \end{array}$$

J
$$\begin{array}{r} 8 \\ +7 \\ \hline \end{array} \qquad \begin{array}{r} 1 \\ +7 \\ \hline \end{array} \qquad \begin{array}{r} 2 \\ +10 \\ \hline \end{array} \qquad \begin{array}{r} 12 \\ -10 \\ \hline \end{array} \qquad \begin{array}{r} 7 \\ -1 \\ \hline \end{array}$$

Skip counting can help you multiply.
Skip count by 2 six times.

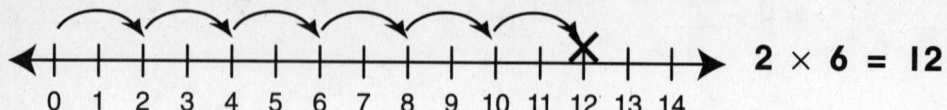

$2 \times 6 = 12$

Skip count by 5 three times.

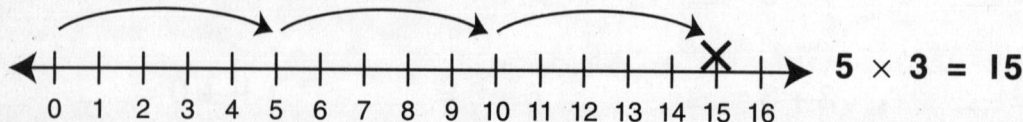

$5 \times 3 = 15$

Skip count by 2s.

A   4 ___   8 ___   ___   14 ___

B   2   4 ___   ___   ___   12 ___

C   8 ___   ___   ___   ___   ___   20

D   6 ___   ___   12 ___   ___   18

Skip count by 5s.

E   15 ___   25 ___   ___   ___   45 ___

F   5 ___   ___   20 ___   ___   35 ___

G   10 ___   20 ___   ___   ___   40

H   20   25 ___   ___   ___   45 ___

Name _____

4 × 2 = 8   5 × 3 = 15

Multiply. Write the product.

A 2 × 3 = _6_   2 × 1 = _2_   2 × 2 = _4_

B 6 × 2 = _12_   8 × 2 = _16_   2 × 0 = _0_

C 1 × 5 = _5_   5 × 7 = _35_   5 × 4 = _20_

D 8 × 5 = _40_   5 × 3 = _15_   5 × 5 = _25_

E

0	9	10	2	9	6
× 5	× 2	× 5	× 7	× 5	× 2

F

8	9	7	4	0	5
× 2	× 5	× 2	× 5	× 2	× 0

G

8	2	5	2	6	2
× 5	× 1	× 4	× 6	× 5	× 5

Name _____

$3 \times 2 = 6$

$4 \times 2 = 8$

Multiply. Write the product.

A $3 \times 6 =$ ___ $3 \times 1 =$ ___ $3 \times 3 =$ ___

B $3 \times 0 =$ ___ $3 \times 8 =$ ___ $3 \times 7 =$ ___

C $4 \times 6 =$ ___ $4 \times 1 =$ ___ $4 \times 5 =$ ___

D $4 \times 7 =$ ___ $4 \times 8 =$ ___ $4 \times 3 =$ ___

E
$$\begin{array}{cccccc}
3 & 5 & 6 & 0 & 9 & 2 \\
\times 3 & \times 4 & \times 3 & \times 4 & \times 3 & \times 4 \\
\end{array}$$

F
$$\begin{array}{cccccc}
1 & 5 & 9 & 8 & 6 & 7 \\
\times 4 & \times 3 & \times 4 & \times 3 & \times 4 & \times 3 \\
\end{array}$$

G
$$\begin{array}{cccccc}
10 & 7 & 0 & 4 & 1 & 8 \\
\times 3 & \times 4 & \times 3 & \times 4 & \times 3 & \times 4 \\
\end{array}$$

Name _____

$$0 \times 6 = 0$$

$$\begin{array}{r} 4 \\ \times\ 1 \\ \hline 4 \end{array}$$

Multiply. Write the product.

A $3 \times 0 = \underline{0}$ $0 \times 8 = \underline{0}$ $0 \times 7 = \underline{0}$

B $0 \times 2 = \underline{0}$ $6 \times 0 = \underline{0}$ $0 \times 5 = \underline{0}$

C $1 \times 9 = \underline{9}$ $6 \times 1 = \underline{6}$ $3 \times 1 = \underline{3}$

D $1 \times 12 = \underline{12}$ $10 \times 1 = \underline{10}$ $1 \times 7 = \underline{7}$

E
$$\begin{array}{r} 1 \\ \times\ 3 \\ \hline 3 \end{array} \qquad \begin{array}{r} 0 \\ \times\ 3 \\ \hline 0 \end{array} \qquad \begin{array}{r} 7 \\ \times\ 1 \\ \hline 7 \end{array} \qquad \begin{array}{r} 2 \\ \times\ 0 \\ \hline 0 \end{array} \qquad \begin{array}{r} 5 \\ \times\ 1 \\ \hline 5 \end{array} \qquad \begin{array}{r} 0 \\ \times\ 8 \\ \hline 0 \end{array}$$

F
$$\begin{array}{r} 3 \\ \times\ 1 \\ \hline 3 \end{array} \qquad \begin{array}{r} 7 \\ \times\ 0 \\ \hline 0 \end{array} \qquad \begin{array}{r} 1 \\ \times\ 5 \\ \hline 5 \end{array} \qquad \begin{array}{r} 9 \\ \times\ 0 \\ \hline 0 \end{array} \qquad \begin{array}{r} 1 \\ \times\ 7 \\ \hline 7 \end{array} \qquad \begin{array}{r} 0 \\ \times\ 6 \\ \hline 0 \end{array}$$

G
$$\begin{array}{r} 1 \\ \times\ 10 \\ \hline 10 \end{array} \qquad \begin{array}{r} 4 \\ \times\ 0 \\ \hline 0 \end{array} \qquad \begin{array}{r} 6 \\ \times\ 1 \\ \hline 6 \end{array} \qquad \begin{array}{r} 0 \\ \times\ 5 \\ \hline 0 \end{array} \qquad \begin{array}{r} 1 \\ \times\ 9 \\ \hline 9 \end{array} \qquad \begin{array}{r} 8 \\ \times\ 0 \\ \hline 0 \end{array}$$

$6 \times 3 = 18$

$$\begin{array}{r} 2 \\ \times\, 8 \\ \hline 16 \end{array}$$

Multiply. Write the product.

A $2 \times 6 =$ ___ $6 \times 4 =$ ___ $5 \times 6 =$ ___

B $6 \times 10 =$ ___ $6 \times 9 =$ ___ $8 \times 6 =$ ___

C $8 \times 9 =$ ___ $8 \times 1 =$ ___ $3 \times 8 =$ ___

D $8 \times 7 =$ ___ $8 \times 4 =$ ___ $5 \times 8 =$ ___

E
$$\begin{array}{r} 8 \\ \times\, 2 \\ \hline \end{array} \quad \begin{array}{r} 7 \\ \times\, 6 \\ \hline \end{array} \quad \begin{array}{r} 7 \\ \times\, 8 \\ \hline \end{array} \quad \begin{array}{r} 10 \\ \times\, 8 \\ \hline \end{array} \quad \begin{array}{r} 0 \\ \times\, 6 \\ \hline \end{array} \quad \begin{array}{r} 4 \\ \times\, 6 \\ \hline \end{array}$$

F
$$\begin{array}{r} 6 \\ \times\, 2 \\ \hline \end{array} \quad \begin{array}{r} 8 \\ \times\, 3 \\ \hline \end{array} \quad \begin{array}{r} 6 \\ \times\, 5 \\ \hline \end{array} \quad \begin{array}{r} 1 \\ \times\, 8 \\ \hline \end{array} \quad \begin{array}{r} 9 \\ \times\, 6 \\ \hline \end{array} \quad \begin{array}{r} 8 \\ \times\, 7 \\ \hline \end{array}$$

G
$$\begin{array}{r} 6 \\ \times\, 1 \\ \hline \end{array} \quad \begin{array}{r} 8 \\ \times\, 5 \\ \hline \end{array} \quad \begin{array}{r} 3 \\ \times\, 6 \\ \hline \end{array} \quad \begin{array}{r} 4 \\ \times\, 8 \\ \hline \end{array} \quad \begin{array}{r} 6 \\ \times\, 8 \\ \hline \end{array} \quad \begin{array}{r} 8 \\ \times\, 9 \\ \hline \end{array}$$

Name _____

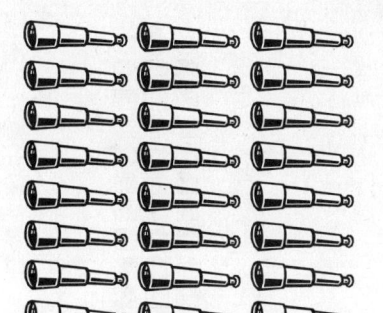

$7 \times 2 = 14$

$$\begin{array}{r} 3 \\ \times 9 \\ \hline 27 \end{array}$$

Multiply. Write the product.

A $7 \times 6 =$ ___ $3 \times 7 =$ ___ $7 \times 9 =$ ___

B $7 \times 8 =$ ___ $7 \times 4 =$ ___ $0 \times 7 =$ ___

C $9 \times 2 =$ ___ $9 \times 4 =$ ___ $6 \times 9 =$ ___

D $5 \times 9 =$ ___ $9 \times 7 =$ ___ $9 \times 9 =$ ___

E
$$\begin{array}{r} 6 \\ \times 7 \\ \hline \end{array} \quad \begin{array}{r} 9 \\ \times 5 \\ \hline \end{array} \quad \begin{array}{r} 7 \\ \times 2 \\ \hline \end{array} \quad \begin{array}{r} 9 \\ \times 0 \\ \hline \end{array} \quad \begin{array}{r} 8 \\ \times 7 \\ \hline \end{array} \quad \begin{array}{r} 9 \\ \times 1 \\ \hline \end{array}$$

F
$$\begin{array}{r} 7 \\ \times 1 \\ \hline \end{array} \quad \begin{array}{r} 8 \\ \times 9 \\ \hline \end{array} \quad \begin{array}{r} 5 \\ \times 7 \\ \hline \end{array} \quad \begin{array}{r} 9 \\ \times 2 \\ \hline \end{array} \quad \begin{array}{r} 9 \\ \times 7 \\ \hline \end{array} \quad \begin{array}{r} 9 \\ \times 9 \\ \hline \end{array}$$

G
$$\begin{array}{r} 4 \\ \times 7 \\ \hline \end{array} \quad \begin{array}{r} 9 \\ \times 3 \\ \hline \end{array} \quad \begin{array}{r} 7 \\ \times 9 \\ \hline \end{array} \quad \begin{array}{r} 6 \\ \times 9 \\ \hline \end{array} \quad \begin{array}{r} 3 \\ \times 7 \\ \hline \end{array} \quad \begin{array}{r} 4 \\ \times 9 \\ \hline \end{array}$$

$10 \times 5 = 50$ $11 \times 3 = 33$ $12 \times 4 = 48$

Multiply. Write the product.

A $10 \times 2 =$ ___ $10 \times 6 =$ ___ $10 \times 4 =$ ___

B $11 \times 5 =$ ___ $11 \times 8 =$ ___ $11 \times 4 =$ ___

C $12 \times 6 =$ ___ $12 \times 5 =$ ___ $12 \times 8 =$ ___

D
$$\begin{array}{r} 4 \\ \times 12 \\ \hline \end{array} \qquad \begin{array}{r} 1 \\ \times 10 \\ \hline \end{array} \qquad \begin{array}{r} 3 \\ \times 11 \\ \hline \end{array} \qquad \begin{array}{r} 9 \\ \times 12 \\ \hline \end{array} \qquad \begin{array}{r} 1 \\ \times 12 \\ \hline \end{array} \qquad \begin{array}{r} 8 \\ \times 10 \\ \hline \end{array}$$

E
$$\begin{array}{r} 2 \\ \times 10 \\ \hline \end{array} \qquad \begin{array}{r} 0 \\ \times 12 \\ \hline \end{array} \qquad \begin{array}{r} 8 \\ \times 11 \\ \hline \end{array} \qquad \begin{array}{r} 1 \\ \times 12 \\ \hline \end{array} \qquad \begin{array}{r} 7 \\ \times 10 \\ \hline \end{array} \qquad \begin{array}{r} 6 \\ \times 11 \\ \hline \end{array}$$

F
$$\begin{array}{r} 5 \\ \times 10 \\ \hline \end{array} \qquad \begin{array}{r} 1 \\ \times 11 \\ \hline \end{array} \qquad \begin{array}{r} 8 \\ \times 12 \\ \hline \end{array} \qquad \begin{array}{r} 0 \\ \times 10 \\ \hline \end{array} \qquad \begin{array}{r} 9 \\ \times 11 \\ \hline \end{array} \qquad \begin{array}{r} 0 \\ \times 11 \\ \hline \end{array}$$

Name_____

A multiplication fact can be turned around. The product is the same.

$6 \times 2 = 12$

$2 \times 6 = 12$

$$\begin{array}{r} 7 \\ \times 3 \\ \hline 21 \end{array} \qquad \begin{array}{r} 3 \\ \times 7 \\ \hline 21 \end{array}$$

Multiply. Write the product.

A $3 \times 6 =$ ___ $6 \times 3 =$ ___   **B** $1 \times 3 =$ ___ $3 \times 1 =$ ___

C $4 \times 5 =$ ___ $5 \times 4 =$ ___   **D** $2 \times 4 =$ ___ $4 \times 2 =$ ___

E $3 \times 5 =$ ___ $5 \times 3 =$ ___   **F** $1 \times 4 =$ ___ $4 \times 1 =$ ___

G $5 \times 6 =$ ___ $6 \times 5 =$ ___   **H** $2 \times 3 =$ ___ $3 \times 2 =$ ___

I $\begin{array}{r} 4 \\ \times 7 \\ \hline \end{array}$ $\begin{array}{r} 7 \\ \times 4 \\ \hline \end{array}$ **J** $\begin{array}{r} 2 \\ \times 7 \\ \hline \end{array}$ $\begin{array}{r} 7 \\ \times 2 \\ \hline \end{array}$ **K** $\begin{array}{r} 3 \\ \times 4 \\ \hline \end{array}$ $\begin{array}{r} 4 \\ \times 3 \\ \hline \end{array}$

L $\begin{array}{r} 4 \\ \times 8 \\ \hline \end{array}$ $\begin{array}{r} 8 \\ \times 4 \\ \hline \end{array}$ **M** $\begin{array}{r} 7 \\ \times 8 \\ \hline \end{array}$ $\begin{array}{r} 8 \\ \times 7 \\ \hline \end{array}$ **N** $\begin{array}{r} 6 \\ \times 8 \\ \hline \end{array}$ $\begin{array}{r} 8 \\ \times 6 \\ \hline \end{array}$

O $\begin{array}{r} 2 \\ \times 5 \\ \hline \end{array}$ $\begin{array}{r} 5 \\ \times 2 \\ \hline \end{array}$ **P** $\begin{array}{r} 7 \\ \times 5 \\ \hline \end{array}$ $\begin{array}{r} 5 \\ \times 7 \\ \hline \end{array}$ **Q** $\begin{array}{r} 8 \\ \times 2 \\ \hline \end{array}$ $\begin{array}{r} 2 \\ \times 8 \\ \hline \end{array}$

A multiplication fact can be turned around. The product is the same.

$$4 \times 12 = 48$$
$$12 \times 4 = 48$$

$$\begin{array}{r} 3 \\ \times 10 \\ \hline 30 \end{array} \qquad \begin{array}{r} 10 \\ \times 3 \\ \hline 30 \end{array}$$

Multiply. Write the product.

A $\quad 4 \times 9 = \underline{\quad} \qquad 9 \times 4 = \underline{\quad}$

B $\quad 2 \times 11 = \underline{\quad} \qquad 11 \times 2 = \underline{\quad}$

C $\quad 8 \times 9 = \underline{\quad} \qquad 9 \times 8 = \underline{\quad}$

D $\quad 6 \times 10 = \underline{\quad} \qquad 10 \times 6 = \underline{\quad}$

E $\quad 4 \times 10 = \underline{\quad} \qquad 10 \times 4 = \underline{\quad}$

F $\quad 2 \times 9 = \underline{\quad} \qquad 9 \times 2 = \underline{\quad}$

G $\quad 3 \times 12 = \underline{\quad} \qquad 12 \times 3 = \underline{\quad}$

H $\quad 7 \times 11 = \underline{\quad} \qquad 11 \times 7 = \underline{\quad}$

I $\quad \begin{array}{r} 5 \\ \times 9 \\ \hline \end{array} \qquad \begin{array}{r} 9 \\ \times 5 \\ \hline \end{array}$
J $\quad \begin{array}{r} 3 \\ \times 9 \\ \hline \end{array} \qquad \begin{array}{r} 9 \\ \times 3 \\ \hline \end{array}$
K $\quad \begin{array}{r} 8 \\ \times 10 \\ \hline \end{array} \qquad \begin{array}{r} 10 \\ \times 8 \\ \hline \end{array}$

L $\quad \begin{array}{r} 3 \\ \times 11 \\ \hline \end{array} \qquad \begin{array}{r} 11 \\ \times 3 \\ \hline \end{array}$
M $\quad \begin{array}{r} 8 \\ \times 12 \\ \hline \end{array} \qquad \begin{array}{r} 12 \\ \times 8 \\ \hline \end{array}$
N $\quad \begin{array}{r} 7 \\ \times 9 \\ \hline \end{array} \qquad \begin{array}{r} 9 \\ \times 7 \\ \hline \end{array}$

O $\quad \begin{array}{r} 2 \\ \times 10 \\ \hline \end{array} \qquad \begin{array}{r} 10 \\ \times 2 \\ \hline \end{array}$
P $\quad \begin{array}{r} 5 \\ \times 11 \\ \hline \end{array} \qquad \begin{array}{r} 11 \\ \times 5 \\ \hline \end{array}$
Q $\quad \begin{array}{r} 6 \\ \times 12 \\ \hline \end{array} \qquad \begin{array}{r} 12 \\ \times 6 \\ \hline \end{array}$

Name_____

Doubling one factor doubles the product.

$2 \times 6 = 12$

so $4 \times 6 = 12 + 12$

$= 24$

4×6

2×6

2×6

Multiply. Write the product.

A $7 \times 3 = $ ___ $7 \times 6 = $ ___

B $5 \times 4 = $ ___ $10 \times 4 = $ ___

C $11 \times 3 = $ ___ $11 \times 6 = $ ___

D $8 \times 2 = $ ___ $8 \times 4 = $ ___

E $5 \times 5 = $ ___ $10 \times 5 = $ ___

F $9 \times 4 = $ ___ $9 \times 8 = $ ___

G $4 \times 3 = $ ___ $8 \times 3 = $ ___

H $6 \times 9 = $ ___ $12 \times 9 = $ ___

I $7 \times 6 = $ ___ $7 \times 12 = $ ___

J $3 \times 5 = $ ___ $6 \times 5 = $ ___

K $5 \times 8 = $ ___ $10 \times 8 = $ ___

L $3 \times 9 = $ ___ $6 \times 9 = $ ___

M $3 \times 8 = $ ___ $6 \times 8 = $ ___

N $2 \times 9 = $ ___ $4 \times 9 = $ ___

O $5 \times 3 = $ ___ $10 \times 3 = $ ___

P $4 \times 8 = $ ___ $8 \times 8 = $ ___

Name_____

Add to find a product.

$$5 \times 3 = 15$$
$$6 \times 3 = 15 + 3 = 18$$

5×3 {baseballs} 6×3

Multiply. Write the product.

A $6 \times 5 =$ ___ $7 \times 5 =$ ___

B $3 \times 9 =$ ___ $4 \times 9 =$ ___

C $10 \times 5 =$ ___ $11 \times 5 =$ ___

D $7 \times 9 =$ ___ $8 \times 9 =$ ___

E $8 \times 5 =$ ___ $9 \times 5 =$ ___

F $6 \times 3 =$ ___ $7 \times 3 =$ ___

G $7 \times 4 =$ ___ $8 \times 4 =$ ___

H $11 \times 7 =$ ___ $12 \times 7 =$ ___

I $6 \times 6 =$ ___ $7 \times 6 =$ ___

J $8 \times 11 =$ ___ $9 \times 11 =$ ___

K $3 \times 12 =$ ___ $4 \times 12 =$ ___

L $5 \times 8 =$ ___ $6 \times 8 =$ ___

M $9 \times 4 =$ ___ $10 \times 4 =$ ___

N $6 \times 7 =$ ___ $7 \times 7 =$ ___

O $3 \times 6 =$ ___ $4 \times 6 =$ ___

P $8 \times 7 =$ ___ $9 \times 7 =$ ___

Q $4 \times 3 =$ ___ $5 \times 3 =$ ___

R $3 \times 9 =$ ___ $4 \times 9 =$ ___

Name_____

Sometimes the first factor is less than the second factor.

$$3 \times 5 = 15$$

$$\begin{array}{r} 5 \\ \times\ 3 \\ \hline 15 \end{array}$$

Multiply. Write the product.

A $6 \times 12 =$ ___ $3 \times 9 =$ ___ $9 \times 10 =$ ___ $5 \times 8 =$ ___

B $7 \times 9 =$ ___ $9 \times 12 =$ ___ $10 \times 11 =$ ___ $6 \times 7 =$ ___

C $8 \times 9 =$ ___ $3 \times 11 =$ ___ $2 \times 10 =$ ___ $4 \times 6 =$ ___

D $5 \times 7 =$ ___ $4 \times 8 =$ ___ $2 \times 9 =$ ___ $6 \times 8 =$ ___

E $3 \times 8 =$ ___ $7 \times 10 =$ ___ $8 \times 11 =$ ___ $11 \times 12 =$ ___

F $7 \times 8 =$ ___ $9 \times 11 =$ ___ $5 \times 9 =$ ___ $4 \times 7 =$ ___

G

6	10	4	11	10	9	12
×5	×4	×3	×6	×8	×2	×5

H

12	12	7	9	6	5	7
×2	×4	×3	×6	×3	×4	×2

Name_____

Sometimes the first factor is greater than the second factor.

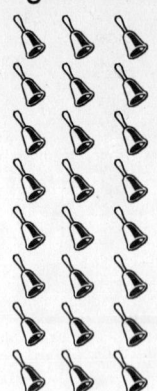

$8 \times 3 = 24$

$$\begin{array}{r} 3 \\ \times 8 \\ \hline 24 \end{array}$$

Multiply. Write the product.

A $6 \times 4 =$ ___ $12 \times 11 =$ ___ $9 \times 4 =$ ___ $11 \times 8 =$ ___

B $10 \times 6 =$ ___ $8 \times 5 =$ ___ $9 \times 8 =$ ___ $11 \times 7 =$ ___

C $12 \times 6 =$ ___ $10 \times 4 =$ ___ $12 \times 5 =$ ___ $9 \times 6 =$ ___

D $12 \times 9 =$ ___ $11 \times 10 =$ ___ $8 \times 7 =$ ___ $10 \times 8 =$ ___

E $7 \times 5 =$ ___ $10 \times 9 =$ ___ $12 \times 4 =$ ___ $11 \times 3 =$ ___

F $10 \times 3 =$ ___ $9 \times 7 =$ ___ $8 \times 6 =$ ___ $9 \times 5 =$ ___

G
$$\begin{array}{r} 7 \\ \times 10 \\ \hline \end{array} \quad \begin{array}{r} 9 \\ \times 11 \\ \hline \end{array} \quad \begin{array}{r} 4 \\ \times 11 \\ \hline \end{array} \quad \begin{array}{r} 7 \\ \times 12 \\ \hline \end{array} \quad \begin{array}{r} 3 \\ \times 9 \\ \hline \end{array} \quad \begin{array}{r} 4 \\ \times 7 \\ \hline \end{array} \quad \begin{array}{r} 5 \\ \times 6 \\ \hline \end{array}$$

H
$$\begin{array}{r} 10 \\ \times 12 \\ \hline \end{array} \quad \begin{array}{r} 3 \\ \times 12 \\ \hline \end{array} \quad \begin{array}{r} 2 \\ \times 11 \\ \hline \end{array} \quad \begin{array}{r} 5 \\ \times 10 \\ \hline \end{array} \quad \begin{array}{r} 6 \\ \times 11 \\ \hline \end{array} \quad \begin{array}{r} 2 \\ \times 12 \\ \hline \end{array} \quad \begin{array}{r} 3 \\ \times 8 \\ \hline \end{array}$$

Name _____

A
10	11	0	12	3	3
×1	×4	×1	×4	×7	×8

B
6	0	6	5	1	9
×6	×8	×7	×7	×4	×0

C
3	11	1	12	1	0
×4	×6	×7	×5	×0	×4

D
8	0	2	12	11	4
×6	×3	×7	×3	×5	×4

E
10	2	10	5	8	6
×5	×8	×6	×0	×4	×4

F
6	9	10	12	7	0
×0	×6	×2	×0	×4	×5

A

5	12	9	10	6	12
×11	×10	×12	×12	×12	×11

B

7	12	11	0	10	11
×12	×12	×11	×12	×10	×12

C

9	11	8	6	12	0
×11	×10	×10	×11	× 9	×11

D

8	0	4	12	7	2
×12	×10	×11	×12	×10	×12

E

0	1	3	9	2	10
× 0	×11	×12	×10	×11	× 7

F

5	4	7	6	8	3
× 12	×10	×11	× 9	× 9	×10

Name_____

A

6	11	1	0	11	12	10
×8	×0	×12	×10	×9	×8	×9

B

2	9	3	4	9	0	4
×0	×7	×11	×12	×9	×9	×8

C

9	5	12	8	4	6	3
×8	×9	×7	×11	×0	×10	×9

D

2	3	1	11	7	7	10
×9	×0	×10	×7	×9	×8	×0

E

8	4	11	0	8	1	6
×7	×9	×8	×7	×0	×8	×1

F

5	0	10	9	12	7	7
×8	×6	×8	×4	×6	×7	×0

G

8	5	0	9	1	4	2
×8	×10	×2	×0	×9	×7	×10

Name _____

A

5	6	11	8	12	10	8
×4	×2	×3	×2	×1	×4	×5

B

8	5	9	9	1	4	7
×3	×3	×1	×2	×6	×6	×5

C

5	11	7	11	9	5	10
×5	×2	×1	×1	×5	×6	×3

D

10	8	2	2	9	6	7
×2	×1	×6	×2	×3	×5	×3

E

7	3	2	3	3	4	1
×6	×3	×1	×1	×6	×3	×1

F

2	5	1	2	3	2	4
×4	×1	×5	×1	×2	×3	×2

G

7	1	5	6	2	4	3
×2	×3	×2	×3	×5	×1	×5

Name _____

Make equal groups. Divide.

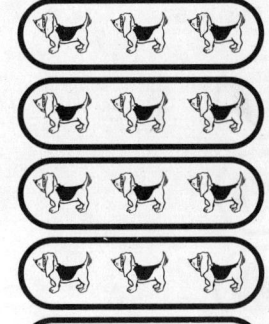

$6 \div 2 = 3$ $15 \div 5 = 3$

Make equal groups. Divide. Write the quotient.

A $2 \div 2 =$ ____ $10 \div 2 =$ ____ $4 \div 2 =$ ____

B $15 \div 5 =$ ____ $0 \div 2 =$ ____ $25 \div 5 =$ ____

C $14 \div 2 =$ ____ $8 \div 2 =$ ____ $20 \div 2 =$ ____

D $20 \div 5 =$ ____ $5 \div 5 =$ ____ $45 \div 5 =$ ____

E $2\overline{)8}$ $2\overline{)16}$ $5\overline{)20}$ $2\overline{)6}$ $5\overline{)50}$ $5\overline{)10}$

F $5\overline{)45}$ $2\overline{)18}$ $2\overline{)12}$ $5\overline{)55}$ $5\overline{)5}$ $2\overline{)22}$

G $2\overline{)20}$ $5\overline{)30}$ $5\overline{)40}$ $2\overline{)24}$ $2\overline{)14}$ $5\overline{)60}$

Name_____

Make equal groups. Divide.

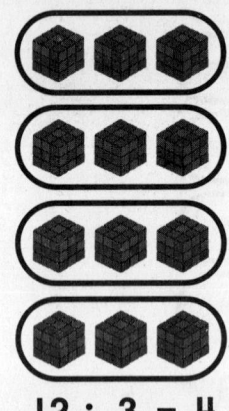

$12 \div 3 = 4$

$8 \div 4 = 2$

Make equal groups. Divide. Write the quotient.

A $27 \div 3 =$ ___ $\qquad$ $28 \div 4 =$ ___ $\qquad$ $4 \div 4 =$ ___

B $20 \div 4 =$ ___ $\qquad$ $21 \div 3 =$ ___ $\qquad$ $3 \div 3 =$ ___

C $12 \div 3 =$ ___ $\qquad$ $36 \div 4 =$ ___ $\qquad$ $16 \div 4 =$ ___

D $9 \div 3 =$ ___ $\qquad$ $8 \div 4 =$ ___ $\qquad$ $15 \div 3 =$ ___

E $4\overline{)28}$ $\quad$ $3\overline{)30}$ $\quad$ $3\overline{)9}$ $\quad$ $4\overline{)8}$ $\quad$ $3\overline{)24}$ $\quad$ $4\overline{)24}$

F $4\overline{)44}$ $\quad$ $4\overline{)36}$ $\quad$ $3\overline{)33}$ $\quad$ $4\overline{)36}$ $\quad$ $3\overline{)21}$ $\quad$ $4\overline{)32}$

G $4\overline{)40}$ $\quad$ $4\overline{)16}$ $\quad$ $3\overline{)36}$ $\quad$ $4\overline{)48}$ $\quad$ $3\overline{)18}$ $\quad$ $3\overline{)12}$

Name_____

Make equal groups. Divide.

$$6 \div 1 = 6$$

$$30 \div 10 = 3$$

Make equal groups. Divide. Write the quotient.

A $90 \div 10 =$ ___ $50 \div 10 =$ ___ $100 \div 10 =$ ___

B $40 \div 10 =$ ___ $60 \div 10 =$ ___ $10 \div 10 =$ ___

C $8 \div 1 =$ ___ $10 \div 1 =$ ___ $4 \div 1 =$ ___

D $5 \div 1 =$ ___ $2 \div 1 =$ ___ $3 \div 1 =$ ___

E $1\overline{)1}$ $10\overline{)30}$ $1\overline{)3}$ $10\overline{)70}$ $1\overline{)10}$ $10\overline{)40}$

F $10\overline{)80}$ $1\overline{)11}$ $10\overline{)60}$ $1\overline{)7}$ $10\overline{)100}$ $1\overline{)2}$

G $10\overline{)110}$ $1\overline{)12}$ $10\overline{)120}$ $1\overline{)6}$ $10\overline{)10}$ $1\overline{)9}$

Name_____

Make equal groups. Divide.

$18 \div 6 = 3$

$21 \div 7 = 3$

Make equal groups. Divide. Write the quotient.

A $42 \div 6 =$ ___ $60 \div 6 =$ ___ $36 \div 6 =$ ___

B $30 \div 6 =$ ___ $6 \div 6 =$ ___ $48 \div 6 =$ ___

C $28 \div 7 =$ ___ $7 \div 7 =$ ___ $56 \div 7 =$ ___

D $14 \div 7 =$ ___ $35 \div 7 =$ ___ $49 \div 7 =$ ___

E $6\overline{)36}$ $7\overline{)63}$ $6\overline{)48}$ $7\overline{)84}$ $6\overline{)18}$ $7\overline{)70}$

F $7\overline{)77}$ $6\overline{)72}$ $4\overline{)36}$ $6\overline{)54}$ $7\overline{)49}$ $6\overline{)60}$

G $7\overline{)14}$ $6\overline{)24}$ $7\overline{)42}$ $6\overline{)6}$ $7\overline{)56}$ $6\overline{)66}$

Make equal groups. Divide.

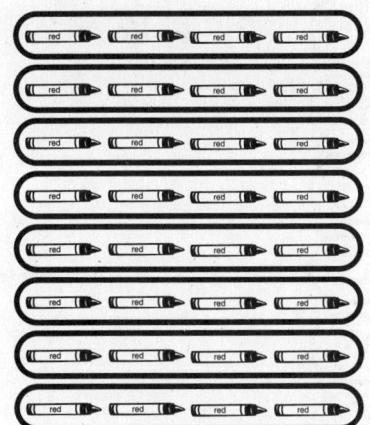

$32 \div 8 = 4$ $27 \div 9 = 3$

Divide. Write the quotient.

A $40 \div 8 = \underline{\quad}$ $80 \div 8 = \underline{\quad}$ $64 \div 8 = \underline{\quad}$

B $16 \div 8 = \underline{\quad}$ $48 \div 8 = \underline{\quad}$ $8 \div 8 = \underline{\quad}$

C $81 \div 9 = \underline{\quad}$ $36 \div 9 = \underline{\quad}$ $99 \div 9 = \underline{\quad}$

D $45 \div 9 = \underline{\quad}$ $18 \div 9 = \underline{\quad}$ $72 \div 9 = \underline{\quad}$

E $8\overline{)32}$ $9\overline{)90}$ $8\overline{)24}$ $9\overline{)36}$ $8\overline{)72}$ $9\overline{)54}$

F $8\overline{)88}$ $9\overline{)9}$ $9\overline{)45}$ $8\overline{)96}$ $8\overline{)48}$ $9\overline{)81}$

G $9\overline{)72}$ $8\overline{)56}$ $9\overline{)63}$ $9\overline{)108}$ $8\overline{)80}$ $8\overline{)40}$

Name _____

Make equal groups. Divide.

 $22 \div 11 = 2$

 $24 \div 12 = 2$

Divide. Write the quotient.

A $11 \div 11 = \underline{\quad}$ $88 \div 11 = \underline{\quad}$ $99 \div 11 = \underline{\quad}$

B $22 \div 11 = \underline{\quad}$ $120 \div 12 = \underline{\quad}$ $144 \div 12 = \underline{\quad}$

C $55 \div 11 = \underline{\quad}$ $44 \div 11 = \underline{\quad}$ $77 \div 11 = \underline{\quad}$

D $48 \div 12 = \underline{\quad}$ $60 \div 12 = \underline{\quad}$ $96 \div 12 = \underline{\quad}$

E $12\overline{)60}$ $12\overline{)108}$ $11\overline{)110}$ $11\overline{)33}$ $12\overline{)84}$ $11\overline{)77}$

F $12\overline{)132}$ $11\overline{)22}$ $12\overline{)96}$ $11\overline{)99}$ $12\overline{)48}$ $11\overline{)121}$

G $11\overline{)55}$ $12\overline{)12}$ $11\overline{)11}$ $12\overline{)120}$ $11\overline{)66}$ $12\overline{)72}$

Name ____________________

Related Division Facts

Related division facts have the same numbers.

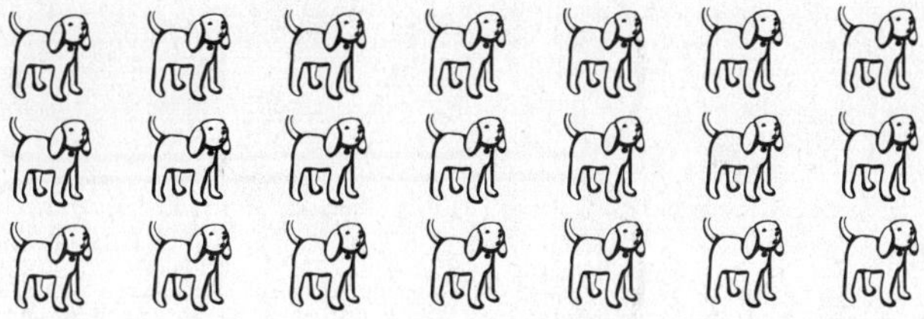

$21 \div 7 = 3$
$21 \div 3 = 7$

Divide. Write the quotient.

A $24 \div 8 =$ ___ $24 \div 3 =$ ___ **B** $30 \div 5 =$ ___ $30 \div 6 =$ ___

C $42 \div 7 =$ ___ $42 \div 6 =$ ___ **D** $24 \div 6 =$ ___ $24 \div 4 =$ ___

E $16 \div 2 =$ ___ $16 \div 8 =$ ___ **F** $18 \div 3 =$ ___ $18 \div 6 =$ ___

G $40 \div 5 =$ ___ $40 \div 8 =$ ___ **H** $15 \div 3 =$ ___ $15 \div 5 =$ ___

I $5\overline{)20}$ $4\overline{)20}$ **J** $8\overline{)32}$ $4\overline{)32}$ **K** $3\overline{)27}$ $9\overline{)27}$

L $5\overline{)45}$ $9\overline{)45}$ **M** $7\overline{)56}$ $8\overline{)56}$ **N** $10\overline{)60}$ $6\overline{)60}$

O $4\overline{)28}$ $7\overline{)48}$ **P** $4\overline{)40}$ $10\overline{)40}$ **Q** $8\overline{)48}$ $6\overline{)48}$

Name _____

The quotient may be greater than the divisor.

$$15 \div 3 = 5$$

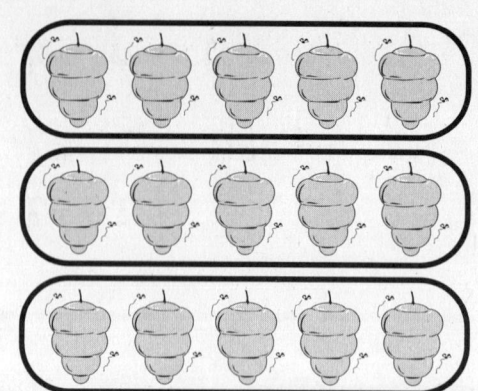

Divide. Write the quotient.

A 72 ÷ 8 = ___   44 ÷ 4 = ___   63 ÷ 7 = ___

B 24 ÷ 3 = ___   12 ÷ 2 = ___   35 ÷ 5 = ___

C 48 ÷ 6 = ___   56 ÷ 7 = ___   70 ÷ 7 = ___

D 36 ÷ 4 = ___   12 ÷ 3 = ___   48 ÷ 6 = ___

E $7\overline{)63}$ $3\overline{)27}$ $4\overline{)48}$ $5\overline{)50}$ $4\overline{)24}$ $6\overline{)42}$

F $6\overline{)54}$ $7\overline{)77}$ $4\overline{)32}$ $2\overline{)12}$ $2\overline{)24}$ $2\overline{)20}$

G $4\overline{)36}$ $3\overline{)15}$ $6\overline{)48}$ $5\overline{)55}$ $5\overline{)40}$ $6\overline{)66}$

Name _____

The quotient may be less than the divisor.

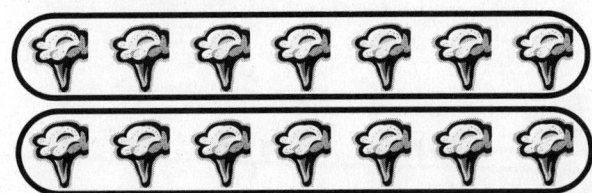

$14 \div 7 = 2$

Divide. Write the quotient.

A $77 \div 11 =$ ___ $45 \div 9 =$ ___ $54 \div 9 =$ ___

B $20 \div 5 =$ ___ $50 \div 10 =$ ___ $21 \div 7 =$ ___

C $18 \div 6 =$ ___ $6 \div 3 =$ ___ $8 \div 4 =$ ___

D $30 \div 6 =$ ___ $70 \div 10 =$ ___ $15 \div 5 =$ ___

E $5\overline{)15}$ $8\overline{)32}$ $11\overline{)99}$ $10\overline{)80}$ $7\overline{)42}$ $6\overline{)12}$

F $7\overline{)14}$ $7\overline{)28}$ $9\overline{)72}$ $12\overline{)84}$ $6\overline{)24}$ $4\overline{)8}$

G $6\overline{)24}$ $7\overline{)56}$ $7\overline{)35}$ $2\overline{)2}$ $5\overline{)10}$ $3\overline{)6}$

A   6)66   4)40   10)50   2)22   5)45   8)8

B   7)35   9)18   1)10   11)55   9)36   9)0

C   10)60   8)40   5)60   10)70   11)22   3)33

D   9)63   8)56   7)70   11)33   8)72   11)0

E   4)44   7)14   5)40   4)36   6)54   9)27

F   5)30   2)20   6)0   6)30   4)28   3)24

A $3\overline{)12}$ $4\overline{)8}$ $2\overline{)2}$ $3\overline{)9}$ $4\overline{)4}$ $5\overline{)10}$

B $11\overline{)121}$ $9\overline{)90}$ $7\overline{)63}$ $9\overline{)108}$ $12\overline{)132}$ $10\overline{)120}$

C $3\overline{)18}$ $6\overline{)18}$ $4\overline{)16}$ $3\overline{)0}$ $1\overline{)1}$ $3\overline{)15}$

D $7\overline{)84}$ $9\overline{)72}$ $9\overline{)99}$ $8\overline{)96}$ $11\overline{)132}$ $12\overline{)96}$

E $8\overline{)80}$ $10\overline{)100}$ $9\overline{)81}$ $11\overline{)55}$ $12\overline{)72}$ $10\overline{)80}$

F $12\overline{)108}$ $12\overline{)144}$ $10\overline{)110}$ $7\overline{)84}$ $11\overline{)77}$ $8\overline{)88}$

Name _____

A $7\overline{)0}$ $9\overline{)9}$ $8\overline{)16}$ $3\overline{)30}$ $10\overline{)10}$ $7\overline{)49}$

B $12\overline{)24}$ $9\overline{)54}$ $11\overline{)88}$ $10\overline{)90}$ $12\overline{)60}$ $8\overline{)48}$

C $9\overline{)45}$ $1\overline{)8}$ $10\overline{)20}$ $7\overline{)28}$ $5\overline{)55}$ $3\overline{)36}$

D $8\overline{)24}$ $11\overline{)11}$ $6\overline{)60}$ $4\overline{)48}$ $10\overline{)0}$ $7\overline{)49}$

E $1\overline{)9}$ $8\overline{)0}$ $2\overline{)24}$ $7\overline{)21}$ $10\overline{)40}$ $8\overline{)32}$

F $3\overline{)27}$ $4\overline{)32}$ $5\overline{)35}$ $6\overline{)36}$ $6\overline{)42}$ $1\overline{)7}$

G $12\overline{)48}$ $1\overline{)11}$ $6\overline{)48}$ $11\overline{)44}$ $7\overline{)42}$ $6\overline{)72}$

Name _____

A $1\overline{)4}$ $5\overline{)0}$ $4\overline{)24}$ $5\overline{)25}$ $2\overline{)18}$ $6\overline{)24}$

B $12\overline{)36}$ $11\overline{)88}$ $9\overline{)72}$ $11\overline{)110}$ $9\overline{)99}$ $6\overline{)54}$

C $4\overline{)20}$ $4\overline{)0}$ $2\overline{)16}$ $5\overline{)20}$ $1\overline{)0}$ $3\overline{)21}$

D $2\overline{)6}$ $3\overline{)6}$ $2\overline{)10}$ $5\overline{)5}$ $3\overline{)3}$ $2\overline{)4}$

E $1\overline{)6}$ $4\overline{)12}$ $1\overline{)2}$ $5\overline{)15}$ $2\overline{)14}$ $1\overline{)5}$

F $11\overline{)66}$ $12\overline{)0}$ $12\overline{)120}$ $12\overline{)84}$ $7\overline{)56}$ $8\overline{)64}$

G $2\overline{)0}$ $1\overline{)3}$ $6\overline{)6}$ $6\overline{)12}$ $10\overline{)30}$ $2\overline{)8}$

Use fact families to multiply and divide.

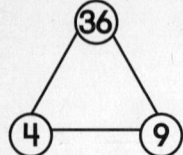

$4 \times 9 = 36$ $36 \div 9 = 4$

$9 \times 4 = 36$ $36 \div 4 = 9$

Multiply and divide.

Complete the facts that make each family.

A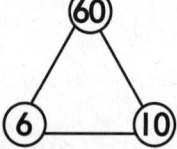

$6 \times 10 =$ ___ $60 \div 6 =$ ___

$10 \times 6 =$ ___ $60 \div 10 =$ ___

B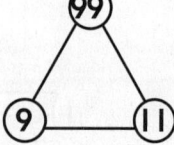

$9 \times 11 =$ ___ $99 \div 9 =$ ___

$11 \times 9 =$ ___ $99 \div 11 =$ ___

C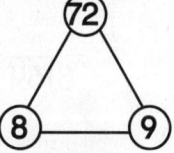

$8 \times 9 =$ ___ $72 \div 8 =$ ___

$9 \times 8 =$ ___ $72 \div 9 =$ ___

D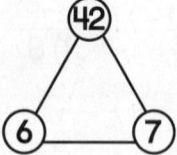

$6 \times 7 =$ ___ $42 \div 6 =$ ___

$7 \times 6 =$ ___ $42 \div 7 =$ ___

E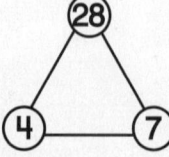

$4 \times 7 =$ ___ $28 \div 4 =$ ___

$7 \times 4 =$ ___ $28 \div 7 =$ ___

Name_____

Use fact families to divide and multiply.

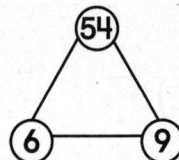

$6 \times 9 = 54$ $54 \div 6 = 9$

$9 \times 6 = 54$ $56 \div 9 = 6$

Multiply or divide.

Complete the facts that make each family.

A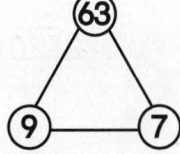

$7 \times 9 = \underline{\quad}$ $63 \div 7 = \underline{\quad}$

$9 \times 7 = \underline{\quad}$ $63 \div 9 = \underline{\quad}$

B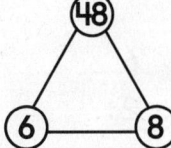

$6 \times 8 = \underline{\quad}$ $48 \div 6 = \underline{\quad}$

$8 \times 6 = \underline{\quad}$ $48 \div 8 = \underline{\quad}$

C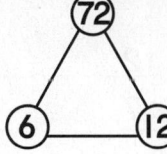

$12 \times 6 = \underline{\quad}$ $72 \div 6 = \underline{\quad}$

$6 \times 12 = \underline{\quad}$ $72 \div 12 = \underline{\quad}$

D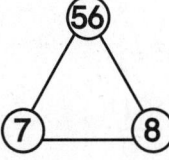

$7 \times 8 = \underline{\quad}$ $56 \div 7 = \underline{\quad}$

$8 \times 7 = \underline{\quad}$ $56 \div 8 = \underline{\quad}$

E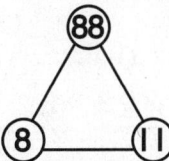

$11 \times 8 = \underline{\quad}$ $88 \div 8 = \underline{\quad}$

$8 \times 11 = \underline{\quad}$ $88 \div 11 = \underline{\quad}$

A $3\overline{)27}$ $4\overline{)32}$ $5\overline{)35}$ $6\overline{)36}$ $6\overline{)42}$ $1\overline{)7}$

B $2\overline{)8}$ $1\overline{)3}$ $1\overline{)6}$ $6\overline{)12}$ $10\overline{)30}$ $2\overline{)0}$

C $7\overline{)7}$ $9\overline{)9}$ $8\overline{)16}$ $3\overline{)30}$ $10\overline{)10}$ $7\overline{)0}$

D $10\overline{)90}$ $8\overline{)48}$ $11\overline{)88}$ $12\overline{)24}$ $12\overline{)60}$ $9\overline{)54}$

E $3\overline{)24}$ $5\overline{)30}$ $5\overline{)0}$ $6\overline{)30}$ $4\overline{)28}$ $2\overline{)20}$

F $11\overline{)121}$ $9\overline{)90}$ $7\overline{)63}$ $9\overline{)108}$ $12\overline{)132}$ $10\overline{)120}$

G

4	12	11	6	8	6	7
$\times 3$	$\times 1$	$\times 3$	$\times 8$	$\times 2$	$\times 2$	$\times 6$

H

4	2	6	5	4	12	8
$\times 6$	$\times 1$	$\times 10$	$\times 9$	$\times 9$	$\times 7$	$\times 11$

I

5	3	2	4	5	3	1
$\times 3$	$\times 6$	$\times 1$	$\times 0$	$\times 5$	$\times 1$	$\times 1$

J

9	1	8	5	2	0	4
$\times 0$	$\times 9$	$\times 8$	$\times 10$	$\times 10$	$\times 2$	$\times 7$

A 5)0̅   1)0̅   4)24̅   5)25̅   2)18̅   6)24̅

B 1)9̅   3)0̅   2)24̅   7)21̅   10)70̅   8)32̅

C 4)16̅   6)18̅   3)15̅   8)0̅   1)1̅   3)18̅

D 10)40̅   8)40̅   5)60̅   10)60̅   11)22̅   3)33̅

E
$$\begin{array}{ccccccc} 5 & 9 & 7 & 10 & 12 & 0 & 7 \\ \times 8 & \times 4 & \times 0 & \times 8 & \times 6 & \times 6 & \times 7 \end{array}$$

F
$$\begin{array}{ccccccc} 5 & 4 & 5 & 2 & 3 & 2 & 1 \\ \times 3 & \times 2 & \times 1 & \times 1 & \times 2 & \times 3 & \times 5 \end{array}$$

G
$$\begin{array}{ccccccc} 2 & 4 & 4 & 3 & 9 & 9 & 0 \\ \times 0 & \times 12 & \times 8 & \times 11 & \times 9 & \times 7 & \times 9 \end{array}$$

A 9)81   10)100   8)80   11)77   12)72   10)80

B 2)22   4)40   10)50   6)66   5)45   8)8

C 11)44   1)11   6)48   12)48   7)42   6)72

D 10)120   9)90   7)63   9)108   12)132   11)121

E 3)36   1)8   10)20   7)28   5)55   9)45

F 9)63   8)56   7)70   11)33   8)72   11)0

G
2	10	9	7	11	3	1
×9	×0	×8	×9	×7	×0	×10

H
10	5	1	3	4	5	2
×2	×1	×5	×2	×2	×3	×1

I
0	6	1	8	8	3	11
×7	×1	×8	×0	×7	×9	×8

J
2	8	2	9	7	6	7
×6	×1	×2	×3	×3	×5	×2

Name _____

A $7\overline{)84}$ $9\overline{)72}$ $9\overline{)99}$ $8\overline{)96}$ $11\overline{)132}$ $12\overline{)96}$

B $12\overline{)0}$ $11\overline{)66}$ $12\overline{)120}$ $12\overline{)36}$ $7\overline{)56}$ $8\overline{)64}$

C $5\overline{)20}$ $4\overline{)0}$ $2\overline{)16}$ $4\overline{)20}$ $1\overline{)4}$ $3\overline{)21}$

D $5\overline{)10}$ $4\overline{)8}$ $2\overline{)2}$ $3\overline{)9}$ $4\overline{)4}$ $3\overline{)12}$

E
$\begin{array}{r}8\\\times5\end{array}$
$\begin{array}{r}11\\\times0\end{array}$
$\begin{array}{r}1\\\times12\end{array}$
$\begin{array}{r}0\\\times10\end{array}$
$\begin{array}{r}11\\\times9\end{array}$
$\begin{array}{r}12\\\times8\end{array}$
$\begin{array}{r}10\\\times9\end{array}$

F
$\begin{array}{r}5\\\times6\end{array}$
$\begin{array}{r}9\\\times2\end{array}$
$\begin{array}{r}6\\\times5\end{array}$
$\begin{array}{r}9\\\times1\end{array}$
$\begin{array}{r}1\\\times6\end{array}$
$\begin{array}{r}7\\\times5\end{array}$
$\begin{array}{r}8\\\times3\end{array}$

G
$\begin{array}{r}9\\\times5\end{array}$
$\begin{array}{r}10\\\times3\end{array}$
$\begin{array}{r}11\\\times2\end{array}$
$\begin{array}{r}5\\\times4\end{array}$
$\begin{array}{r}7\\\times8\end{array}$
$\begin{array}{r}7\\\times1\end{array}$
$\begin{array}{r}11\\\times1\end{array}$

A $10\overline{)110}$ $12\overline{)144}$ $12\overline{)108}$ $7\overline{)84}$ $11\overline{)55}$ $8\overline{)88}$

B $8\overline{)24}$ $11\overline{)11}$ $6\overline{)60}$ $4\overline{)48}$ $10\overline{)0}$ $7\overline{)49}$

C $12\overline{)84}$ $11\overline{)88}$ $9\overline{)72}$ $11\overline{)110}$ $9\overline{)99}$ $6\overline{)54}$

D
$\begin{array}{r} 6 \\ \times 6 \\ \hline \end{array}$
$\begin{array}{r} 6 \\ \times 7 \\ \hline \end{array}$
$\begin{array}{r} 1 \\ \times 4 \\ \hline \end{array}$
$\begin{array}{r} 9 \\ \times 0 \\ \hline \end{array}$
$\begin{array}{r} 5 \\ \times 7 \\ \hline \end{array}$
$\begin{array}{r} 0 \\ \times 8 \\ \hline \end{array}$

E
$\begin{array}{r} 0 \\ \times 5 \\ \hline \end{array}$
$\begin{array}{r} 0 \\ \times 10 \\ \hline \end{array}$
$\begin{array}{r} 2 \\ \times 11 \\ \hline \end{array}$
$\begin{array}{r} 12 \\ \times 12 \\ \hline \end{array}$
$\begin{array}{r} 5 \\ \times 12 \\ \hline \end{array}$
$\begin{array}{r} 8 \\ \times 12 \\ \hline \end{array}$

F
$\begin{array}{r} 3 \\ \times 4 \\ \hline \end{array}$
$\begin{array}{r} 1 \\ \times 7 \\ \hline \end{array}$
$\begin{array}{r} 6 \\ \times 0 \\ \hline \end{array}$
$\begin{array}{r} 0 \\ \times 3 \\ \hline \end{array}$
$\begin{array}{r} 12 \\ \times 0 \\ \hline \end{array}$
$\begin{array}{r} 11 \\ \times 6 \\ \hline \end{array}$

G
$\begin{array}{r} 1 \\ \times 0 \\ \hline \end{array}$
$\begin{array}{r} 10 \\ \times 2 \\ \hline \end{array}$
$\begin{array}{r} 7 \\ \times 4 \\ \hline \end{array}$
$\begin{array}{r} 0 \\ \times 4 \\ \hline \end{array}$
$\begin{array}{r} 10 \\ \times 1 \\ \hline \end{array}$
$\begin{array}{r} 8 \\ \times 6 \\ \hline \end{array}$

H
$\begin{array}{r} 12 \\ \times 5 \\ \hline \end{array}$
$\begin{array}{r} 0 \\ \times 1 \\ \hline \end{array}$
$\begin{array}{r} 3 \\ \times 7 \\ \hline \end{array}$
$\begin{array}{r} 3 \\ \times 4 \\ \hline \end{array}$
$\begin{array}{r} 12 \\ \times 4 \\ \hline \end{array}$
$\begin{array}{r} 11 \\ \times 4 \\ \hline \end{array}$

I
$\begin{array}{r} 9 \\ \times 6 \\ \hline \end{array}$
$\begin{array}{r} 2 \\ \times 7 \\ \hline \end{array}$
$\begin{array}{r} 11 \\ \times 5 \\ \hline \end{array}$
$\begin{array}{r} 4 \\ \times 4 \\ \hline \end{array}$
$\begin{array}{r} 12 \\ \times 3 \\ \hline \end{array}$
$\begin{array}{r} 0 \\ \times 0 \\ \hline \end{array}$

A 1)‾10̄   9)‾18̄   7)‾35̄   11)‾55̄   9)‾36̄   9)‾0̄

B 3)‾15̄   6)‾18̄   4)‾16̄   3)‾0̄   1)‾1̄   3)‾18̄

C 9)‾27̄   7)‾14̄   5)‾40̄   4)‾36̄   6)‾54̄   4)‾44̄

D
$$\begin{array}{r}12\\ \times 11\\\hline\end{array}\quad\begin{array}{r}10\\ \times 12\\\hline\end{array}\quad\begin{array}{r}6\\ \times 12\\\hline\end{array}\quad\begin{array}{r}11\\ \times 11\\\hline\end{array}\quad\begin{array}{r}9\\ \times 12\\\hline\end{array}\quad\begin{array}{r}12\\ \times 10\\\hline\end{array}$$

E
$$\begin{array}{r}0\\ \times 11\\\hline\end{array}\quad\begin{array}{r}7\\ \times 12\\\hline\end{array}\quad\begin{array}{r}12\\ \times 9\\\hline\end{array}\quad\begin{array}{r}8\\ \times 10\\\hline\end{array}\quad\begin{array}{r}6\\ \times 11\\\hline\end{array}\quad\begin{array}{r}11\\ \times 10\\\hline\end{array}$$

F
$$\begin{array}{r}3\\ \times 10\\\hline\end{array}\quad\begin{array}{r}6\\ \times 9\\\hline\end{array}\quad\begin{array}{r}1\\ \times 0\\\hline\end{array}\quad\begin{array}{r}3\\ \times 12\\\hline\end{array}\quad\begin{array}{r}7\\ \times 11\\\hline\end{array}\quad\begin{array}{r}4\\ \times 10\\\hline\end{array}$$

G
$$\begin{array}{r}9\\ \times 11\\\hline\end{array}\quad\begin{array}{r}0\\ \times 12\\\hline\end{array}\quad\begin{array}{r}10\\ \times 10\\\hline\end{array}\quad\begin{array}{r}11\\ \times 12\\\hline\end{array}\quad\begin{array}{r}5\\ \times 11\\\hline\end{array}\quad\begin{array}{r}12\\ \times 12\\\hline\end{array}$$

H
$$\begin{array}{r}2\\ \times 12\\\hline\end{array}\quad\begin{array}{r}12\\ \times 11\\\hline\end{array}\quad\begin{array}{r}7\\ \times 10\\\hline\end{array}\quad\begin{array}{r}8\\ \times 12\\\hline\end{array}\quad\begin{array}{r}4\\ \times 11\\\hline\end{array}\quad\begin{array}{r}0\\ \times 10\\\hline\end{array}$$

I
$$\begin{array}{r}0\\ \times 5\\\hline\end{array}\quad\begin{array}{r}8\\ \times 9\\\hline\end{array}\quad\begin{array}{r}7\\ \times 4\\\hline\end{array}\quad\begin{array}{r}6\\ \times 6\\\hline\end{array}\quad\begin{array}{r}10\\ \times 1\\\hline\end{array}\quad\begin{array}{r}10\\ \times 2\\\hline\end{array}$$

Multiply or divide to find customary length equivalents.

Customary Units of Length
I foot (ft) = 12 inches (in.)
I yard (yd) = 36 inches
I yard = 3 feet
I mile (mi) = 5,280 feet
I mile = 1,760 yd

2 ft = _____ in.
Think: I ft = 12 in.
$2 \times 12 = 24$
2 ft = 24 in.

Complete. Find the equivalent measure of customary length.

A   4 ft = _____ in.   3 yd = _____ ft   2 mi. = _____ ft

B   7 yd = _____ in.   5 mi = _____ yd   11 ft = _____ in.

C   72 in = _____ ft   15 ft = _____ yd   3,520 yd = _____ mi

D   144 in. = _____ yd   15,840 ft = _____ mi   96 in. = _____ ft

E $3\frac{1}{2}$ ft = _____ in. $1\frac{1}{4}$ mi = _____ ft $5\frac{2}{3}$ yd = _____ ft

F   20 in. = _____ ft _____ in.   20 ft = _____ yd _____ ft   80 in. = _____ yd _____ ft

G   5 ft. 8 in. = _____ in.   7 yd 2 ft = _____ ft   5 yd. 1 ft = _____

Name_____

Multiply or divide to find customary weight equivalents.

Customary Units of Weight
I pound (lb) = 16 ounces (oz)
I ton (T) = 2,000 pounds

48 oz = _____ lb
Think: 16 oz = 1 lb
48 ÷ 16 = 3
48 oz = 3 lb

Complete. Find the equivalent measure of customary weight.

A 2 lb = _____ oz   6 T = _____ lb   9 lb = _____ oz

B 10 T = _____ lb   12 lb = _____ oz   25 T = _____ lb

C 64 oz = _____ lb   8,000 lb = _____ T   160 oz = _____ lb

D 28,000 lb = _____ T   1,600 oz = _____ lb   32,000 oz = _____ T

E $5\frac{1}{2}$ lb = _____ oz $3\frac{1}{4}$ T = _____ lb   32 T = _____ lb

F 7 lb 9 oz = _____ oz   5 T 350 lb = _____ lb   1 T 500 lb = _____ oz

G 68 oz = _____ lb _____ oz 6,250 lb = _____ T _____ lb 50 oz = _____ lb _____ oz

Multiply or divide to find customary length equivalents.

Customary Units of Capacity
I cup = 8 fluid ounces (fl oz)
I pint (pt) = 2 cups
I quart = 4 pints
I quart = 4 cups
I gallon (gal) = 4 quarts

2 ft = _____ in.
Think: I ft = 12 in.
$2 \times 12 = 24$
2 ft = 24 in.

Complete. Find the equivalent measure of customary length.

A   3 qt = _____ c   5 gal = _____ qt   7 pt = _____ c

B   6 c = _____ fl oz   12 qt = _____ pt   5 pt = _____ fl oz

C   16 c = _____ qt   96 c = _____ gal   64 fl oz = _____ c

D   160 fl oz = _____ qt   72 pt = _____ qt   144 qt = _____ gal

E $2\frac{1}{2}$ qt = _____ fl oz $1\frac{5}{8}$ c = _____ fl oz $3\frac{3}{4}$ gal = _____ qt

F   20 c = _____ qt   28 fl oz = _____ c   17 qt = _____ gal

G   7 c 3 fl oz = _____ fl oz   9 qt 2 pt = _____ pt   3 gal 2 qt= _____ c

Name_____

Multiply or divide to find customary weight equivalents.

Units of Time
I minute (min) = 60 seconds (s)
I hour (h) = 60 minutes
I day (d) = 24 hours
I week (wk) = 7 days
I year (y) = 12 months (mo)
I year = 365 days

30 mo = _____
Think: 12 mo = 1y
$30 \div 12 = 2\ R\ 6$ $\frac{6}{12} = \frac{1}{2}$
30 mo = $2\frac{1}{2}$ y

Complete. Find the equivalent measure of time.

A 5 h = _____ min   7 min = _____ sec   8 wk = _____ d

B 9 y = _____ mo   2 y = _____ d   3 d = _____ h

C 240 min = _____ h   660 sec = _____ min   120 h = _____ d

D 156 mo = _____ y   84 d = _____ wk   240 mo = _____ y

E $5\frac{1}{2}$ h = _____ min $3\frac{2}{3}$ min = _____ sec $6\frac{3}{4}$ = _____ oz

F 45 sec = _____ min   45 mo = _____ y   23 d = _____ wk

G I y 200 d = _____ d   4 d 7 h = _____ h   7 wk 2 d = _____ d

Multiply or divide to find customary length equivalents.

Metric Units of Length
I centimeter (cm) = 10 millimeters
I decimeter (dm) = 10 centimeters
I meter (m) = 100 centimeters
I kilometer (km) = 1,000 meters

5.3 m = _____ cm
Think: I m = 100 cm
$5.3 \times 100 = 530$
5.3 m = 530 cm

Complete. Find the equivalent measure of metric length.

A   4 cm = _____ mm   9 km = _____ m   15 m = _____ cm

B   2.7 dm = _____ cm   0.84 m = _____ cm   1.09 m = _____ mm

C   79 c = _____ qt   1,500 m = _____ km   250 mm = _____ cm

D   805 mm = _____ cm   4.7 cm = _____ dm   570 mm = _____ m

E   528 dm = _____ mm   0.082 = _____ m   1.007 = _____ mm

F   71.76 mm = _____ cm   45,000 = _____ km   0.84 km = _____ dm

G   9 dm 7 cm = _____ mm 7 m 2 dm = _____ mm   3 km 100 m = _____ cm

Multiply or divide by powers of ten to find metric mass

Metric Units of Mass
I centigram (cg) = 10 milligrams (mg)
I decigram (dg) = 10 centigrams
I gram (g) = 100 centigrams
I kilogram (kg) = 1,000 grams
I metric ton (t) = 1,000 kilograms

5,400 kg = _____ t
Think: 1,000 kg = 1 t
5,400 ÷ 1,000 = 5.4
5,400 kg = 5.4 t

Complete. Find the equivalent metric mass

A 45 g = _____ cg   9 kg = _____ g   180 dg = _____ cg

B 3.03 kg = _____ g   0.6 t = _____ kg   0.58 dg = _____ mg

C 1.26 mg = _____ g   500 cg = _____ g   64 g = _____ kg

D 2.4 mg = _____ dg   8,000 cg = _____ kg   703 dg = _____ kg

Name_____

Multiply or divide to find customary length equivalents.

Metric Units of Capacity
I centiliter (cL) = 10 millimeters (mL)
I deciliter (dL) = 10 centiliters
I liter (L) = 100 centiliters
I kilometer (kL) = 1,000 liters

26 cL = _____ L.
Think: I L = 100 cL.
26 ÷ 100 = 0.26
26 cL = 0.26L

Complete. Find the equivalent measure of metric length.

A 8 L = _____ dL   12 kL = _____ L   65 L = _____ cL

B 0.7 dL = _____ mL   3.06 L = _____ mL   20.9 cL = _____ mL

C 95 mL = _____ L   650 L = _____ kL   0.03 cL = _____ L

D 506 mL = _____ dL   15,000 cL = _____ kL   444 mL = _____ dL

E 4.006 L = _____ cL   7,250 dL = _____ kL   32,000 cL= _____ kL

F 9.13 L = _____ dL   0.008 kL = _____ mL   16,000 mL= _____ L

G 9 dL 6 cL = _____ mL   7 L 2 dL = _____ cL   6 L 50 cL = _____ mL

Name _____

To find the fraction equivalent to a decimal, rewrite the decimal as a fraction, and simplify.

$0.8 =$ _____ Think: 8 tenths or $\frac{8}{10}$

$0.8 = \frac{8}{10}$

$\frac{8}{10} = \frac{4}{5}$

$0.8 = \frac{4}{5}$

To find the deimal equivalent to a fraction, divide the numerator by the denominator.

$\frac{3}{5} =$ _____

$$5\overline{)3.0}^{\,0.6}$$

$\frac{3}{5} = 0.6$

Complete. Find the fraction equivalent to the decimal or the decimal equivalent to the fraction.

A $0.4 =$ _____ $0.15 =$ _____ $0.35 =$ _____

B $0.82 =$ _____ $0.45 =$ _____ $0.08 =$ _____

C $0.17 =$ _____ $0.78 =$ _____ $0.6 =$ _____

D $\frac{2}{5} =$ _____ $\frac{3}{10} =$ _____ $\frac{17}{20} =$ _____

E $\frac{3}{4} =$ _____ $\frac{7}{10} =$ _____ $\frac{27}{50} =$ _____

F $\frac{9}{10} =$ _____ $\frac{3}{5} =$ _____ $\frac{7}{8} =$ _____

G $\frac{4}{5} =$ _____ $\frac{5}{8} =$ _____ $\frac{18}{25} =$ _____

To find the percent equivalent to a decimal, move the decimal point two places to the right and write the percent symbol.

$0.07 =$ _____ %

         0.07   0 0 7

0.07 7

$0.07 = 7\%$

To find the decimal equivalent, drop the pecent symbol and move the decimal point two places to the left.

$49\% =$ _____

              0 4 9.   0.4.9.

   49   0.49

$49\% = 0.49$

Complete. Find the percent equivalent to the decimal or the decimal equivalent to the percent.

A $0.32 =$ _____ % $0.75 =$ _____ % $0.6 =$ _____ %

B $0.09 =$ _____ % $1.2 =$ _____ % $0.004 =$ _____ %

C $3.04 =$ _____ % $0.335 =$ _____ % $2.035 =$ _____ %

D $39\% =$ _____ $11\% =$ _____ $2\% =$ _____

E $6.4\% =$ _____ $0.3\% =$ _____ $200\% =$ _____

F $0.81\% =$ _____ $76.2\% =$ _____ $0.5\% =$ _____

G $350\% =$ _____ $20.07\% =$ _____ $198.4\% =$ _____

Name_____

To find the percent equivalent to a fraction, write the fraction as a decimal. Then write the decimal as a percent.

$\frac{3}{4} = $ _____ %

$\frac{3}{4} = 0.75$

$0.75 = 75\%$

$\frac{3}{4} = 75\%$

To find the fraction equivalent to a percent, write the percent as a decimal. Then write the decimal as a fraction, and simplify.

$48\% = $ _____

$48\% = 0.48$

$0.48 = \frac{48}{100} = \frac{12}{25}$

$48\% = \frac{12}{25}$

Complete. Find the percent equivalent to the fraction or the fraction equivalent to the percent.

A $\frac{3}{5} = $ _____ % $\frac{11}{50} = $ _____ % $\frac{3}{100} = $ _____ %

B $\frac{7}{10} = $ _____ % $\frac{2}{25} = $ _____ % $\frac{1}{2} = $ _____ %

C $\frac{3}{4} = $ _____ % $\frac{13}{20} = $ _____ % $\frac{7}{8} = $ _____ %

D $80\% = $ _____ $36\% = $ _____ $9\% = $ _____

E $62\% = $ _____ $6\% = $ _____ $37.5\% = $ _____

F $92\% = $ _____ $10\% = $ _____ $44\% = $ _____

G $62.5\% = $ _____ $55\% = $ _____ $120\% = $ _____

Use formulas to find perimeter, circumference, area, or volume of plane and solid figures.

Geometric Formulas	
Perimeter of a rectangle: $2(l+w)$	Circumference of a circle: πd, $2\pi r$
Perimeter of a square: $4s$	Area of a circle: πr^2
Area of a rectangle: $l \times w$	Volume of rectangular prism: $l \times w \times h$
Area of a Square: s^2	
Area of a parallelogram: $b \times h$	Volume of any prism: $B \times h$
Area of a trangle: $\frac{1}{2}(b \times h)$	Volume of cyclinder: $\pi r^2 h$

Complete. Find the equivalent measure of metric length.

A Figure: Rectangle
length = 6 in., width = 4 in.

Perimeter = _____

Figure: Square
side = 8 cm

Area = _____

B Figure: Triangle
base = 9 feet, height = 7 ft

Area = _____

Figure: Parallelogram
base = 12.5 m, height = 4m

Area = _____

C Figure: Circle
diameter = 6 in.

Area = _____

Figure: Circle
radius = 28 mm

Circumference = _____

D Figure: Prism
area of base = 49 in.²
height = 12 in.

Volume = _____

Figure: Cylinder
radius = 7 m, height = 20 m

Volume = _____

E Figure: Rectangle Prism
length = 1.6 cm, width = 3.1 cm
height = 2.5

Volume = _____

Figure: Square
side = 25 ft

Perimeter = _____

F Figure: Rectangle
length = 20 m, width = 68 m

Area = _____

Figure: Circle
radius = $3\frac{1}{2}$ yd

Circumference = _____

+	1	2	3	4	5	6	7	8	9	10
1										
2										
3										
4										
5										
6										
7										
8										
9										
10										

×	0	1	2	3	4	5	6	7	8	9	10	11	12
0	0	0	0	0	0	0	0	0	0	0	0	0	0
1	0	1	2	3	4	5	6	7	8	9	10	11	12
2	0	2	4	6	8	10	12	14	16	18	20	22	24
3	0	3	6	9	12	15	18	21	24	27	30	33	36
4	0	4	8	12	16	20	24	28	32	36	40	44	48
5	0	5	10	15	20	25	30	35	40	45	50	55	60
6	0	6	12	18	24	30	36	42	48	54	60	66	72
7	0	7	14	21	28	35	42	49	56	63	70	77	84
8	0	8	16	24	32	40	48	56	64	72	80	88	96
9	0	9	18	27	36	45	54	63	42	81	90	99	108
10	0	10	20	30	40	50	60	70	80	90	100	110	120
11	0	11	22	33	44	55	66	77	88	99	110	121	132
12	0	12	24	36	48	60	72	84	96	108	120	132	144